No He's Not a Monkey, He's an Ape and He's My Son

Also by Hester Mundis

Jessica's Wife

No He's Not a Monkey,

Photo: David Sagarin

by Hester Mundis

He's an Ape and He's My Son

CROWN PUBLISHERS, INC., NEW YORK

A letter to Tom and Charlie
written with love for
Thelma Williams
Nick Carrado
and
Ken Green

Grateful acknowledgment is made to the following authors, whose books not only enlightened and pleasured me but comforted me as well.

Bourne, Geoffrey H. *The Ape People.* New York: G.P. Putnam's Sons, 1971.
Fisher, James. *Zoos of the World: The Story of Animals in Captivity.* Garden City, N.Y.: Natural History Press, 1967.
Goodall, Jane Van Lawick. *In the Shadow of Man.* Boston: Houghton Mifflin Co., 1971.
Hahn, Emily. *Animal Gardens.* Garden City, N.Y.: Doubleday & Co., 1967.
______. *On the Side of the Apes.* New York: Thomas Y. Crowell Co., 1971.
Hayes, Cathy. *The Ape in Our House.* New York: Harper & Brothers, 1951.
Hunt, John. *A World Full of Animals.* New York: David McKay Co., 1969.
Morris, Desmond, ed. *Primate Ethology.* Chicago: Aldine Publishing Co., 1967.
Morris, Ramona and Desmond. *Men and Apes.* New York: McGraw-Hill Book Co., 1966.
Mundis, Jerrold. *The Guard Dog.* New York: David McKay Co., 1970.
Oberjohann, Heinrich. *My Friend the Chimpanzee.* Translated by Monica Brooksbank. London: Robert Hale, 1957.
Reynolds, Vernon. *The Apes.* New York: E.P. Dutton & Co., 1967.
Spock, Benjamin. *Baby and Child Care.* New York: Pocket Books, 1968.
Trefflich, Henry, and Anthony, Edward. *Jungle for Sale.* New York: Hawthorn Books, 1967.

Printed in the United States of America

Published simultaneously in Canada by General Publishing Company Limited

Library of Congress Cataloging in Publication Data

Mundis, Hester.
No he's not a monkey, he's an ape and he's my son.

1. Chimpanzees—Legends and stories.
I. Title.
QL795.C57M86 818'.5'407 76-12527
ISBN 0-517-52669-7

Contents

CHAPTER 1

* *A Trip Downtown* *

Being at home on a rainy weekend with your eight-year-old son is not the worst thing in the world. It's not the best, either. On one particularly bleak Saturday in March my husband Jerry and I were seriously considering locking ourselves in our bedroom and pretending we couldn't get out. All hope of sending Shep somewhere else to play had been snuffed out by a mushroom-soup mist that obliterated Manhattan's upper West Side (probably the rest of the city too, and maybe even the state, but who cared about that?), and for over an hour we had been assaulted by the most insidious weapons known to modern child—words. *Lots* of them, specifically those infamous rainy-day deadlies that can turn even the most placid parent into a banshee: "There's nothing to do!" . . . "I did that already!" . . . "I don't want to!" . . . "It's broken!" . . . "I hate it!" In the field of word-to word combat, Shep was a black belt.

To make things worse, Jerry and I were wrestling with some heavy parental guilt. Since he had been working obsessively on a book for several months and I had spent a lot of extra hours at the office, Shep had really gotten the short end of the family-togetherness stick. Obviously, we were depriving him of . . . something—fond memories? a firm nuclear family? a healthy basis for future marital happiness? We had no choice. The time had come to right our wrongs, assuage our guilt, and save our sanity. We told Shep to round up his friend Roger and we would all do something special, together. He refused to do so until we assured him that it would not be another puppet show or museum trip, or any activity recommended for kids by *The New York Times*. Shep was eight going on forty-nine.

A friend of ours, whose favorite pastime was keeping abreast of bizarre activities in town (symposiums on the metaphysics of Laurel and Hardy, local celebrations of obscure Serbo-Croatian holidays, and the like), had once mentioned an exotic-pet emporium called Trefflich's. It was a place where wild baby animals were for touching, where children could act out their *Born Free* fantasies, where parents did not have to carry the kids on their backs or buy balloons. It sounded perfect. We headed downtown.

Once upon Fulton Street there *had* been a Trefflich emporium of exotic fauna, but time and taxes take their toll, and when we arrived at the new address listed in the phone book we found a small, rapidly deteriorating pet shop with all the exotic atmosphere of a railroad freight station, and what appeared to be a very slim array of unusual animals. The minute we got there something told me we shouldn't have come. It was Shep.

"We shouldn't have come, Mom," he said. He was a strong believer in first impressions. "There's nothing to see."

"Of course there is," I said without conviction. There had to be something in that cluttered feed-and-cage-filled front room to spark excitement. I looked around. There were a few finger-smudged vivariums stocked with sleeping snakes, and an aging parrot that could probably talk but didn't, and a small glass room that housed two puppies, a cat, and three monkeys.

"Oh, wow, look at the monkeys," I said, nudging the boys toward the room.

"They can't go in there," a shop attendant said, slamming the door. It boded ill.

Jerry beckoned to us from a doorway at the rear of the shop. "They're back here."

And they were. Cages of monkeys upon monkeys, mynah birds, a pygmy hippopotamus that we were told was very affectionate and made a great pet for people with large bathtubs, a pair of lion cubs guaranteed to be unmanageable in two years (a fact that to my consternation seemed to increase Jerry's interest in them), and Heinrich, an ape that could have walked off with the lead in any road-show company of "King Kong." Easily.

"Boys, come look at the gorilla," I said, six short words that got me a swift jab in the ribs from Jerry, who as a serious writer had nothing but wild disdain for blatant displays of ignorance, particularly when they were public. Heinrich, he told me through clenched teeth, was a chimpanzee.

It was like learning all over again that there was no Santa Claus. Poof! There went my whole childhood Sheena, Queen of the Jungle fantasy, in which I, dauntless, daring, and stacked in my leopard-skin sarong, would swing through the trees with a loving chimp astride my back. I took another look at Heinrich. So long sarong.

"He's nine years old," the attendant said. "Watch this." He turned toward the cage. "What do you think of that, Heinrich?"

Heinrich gave a juicy Bronx cheer.

Jerry seemed enchanted. "Is he hard to handle?" he asked.

"Naw, not really." The attendant proceeded to unlock Heinrich's cage. "Come on out, boy." He extended his hand toward the ape. I wondered whether it was a friendly gesture or a sacrificial offering. Heinrich stood about five feet tall and weighed in at approximately one hundred thirty pounds.

"Has the strength of six men," the attendant said proudly. "And he's toilet trained."

I didn't grasp the connection, but I nodded appreciatively. Anyone that close to Heinrich would nod appreciatively. "What happened to his front teeth?" I asked. There were formidable vacancies in the front of his mouth.

"Pulled. He was a show-biz chimp. No sense taking chances, you know. Wanna see him use the toilet?"

"Well, not really," I said, but too late to stop the attendant, to say nothing of Heinrich.

This time Jerry was visibly impressed. "That's great!" he said.

"Very nice," I murmured.

"What do you think of that?" the attendant said.

Heinrich gave another Bronx cheer.

Shep and Roger missed all of this. They were in the front of the shop, Roger debating whether or not to blow his dollar on some hamster food and Shep squeezing every doggie squeak-toy in the store. They were bored and showing it. It was time to leave.

"Pssssst." One of the animal handlers cocked his head significantly. "Wanna see something?" I froze. (French animal postcards?) He pointed to a kitten-carrying case on the floor behind the counter.

Jerry and I leaned over to look.

The die was cast.

The handler opened the lid and there, lying on a blanket of shredded newspapers, was an adorable and frightened baby chimpanzee. Dark chocolate eyes were set in a light mocha face that was as soft as doeskin, and on his chin was a powder-white fuzz of a beard. His hair was silky and black and parted in the center of his head, bristling out at the sides around two outrageously comic big ears. Something this cute could not be real. It was undoubtedly a very ingenious battery-operated toy. Somewhere on his underside there had to be a tag that said "Made in Japan." But suddenly I was holding him and there was no tag in sight. In one magic moment he threw his arms around my neck, thoroughly wet my coat, and, though I did not know it then, totally annihilated a lifetime of rationality and logic.

"Cute little guy, isn't he?" the handler said. "Has a really good temperament, too."

"How can you tell?" Jerry asked.

"Easy," the handler said.

"How old is he?"

"Few months."

"Difficult to raise?"

"Naw. All you need's a playpen with a top to keep him in."

"What about food?"

"Soon as he's off his formula, just give him some apples, bananas, and lettuce, once a day."

"Hmmmmmm," said Jerry.

I didn't like the sound of it.

"How much?" Jerry asked.

"Well . . ." the handler drawled, "chimps run between eight hundred and a thousand dollars. Like I said, this one is really a good chimp. He's a thousand."

I liked the sound of that even less, but I was reasonably confident it would put an end to Jerry's questions. At that particular juncture in our far-from-solvent lives we were about five thousand dollars behind on an impressive amalgam of debts.

"I think the chimp wants to go back into his box," I lied. As much as I wanted to hold him (and that was quite a bit, since I had always loved those hairy anthropoids with a passion I would never confide even to my analyst), I felt that the sooner he was out of my arms the saner Jerry would be.

"Give him a banana," one of the attendants said. He handed me what was left of the one he'd been eating.

At this point, the chimp was doing an awfully good Boston-strangler imitation around my neck. Every time I loosened one of his arms, the other tightened. "I . . . um . . . think he's frightened," I said.

"Just give 'im the banana, he'll be fine."

Raising my voice an octave and holding the banana over my left shoulder so the chimp, who had already tasted my ear lobe and was now mouthing my clavicle, could see it, I said, "Would you like a 'nana, boy?" My answer was two of the softest, gentlest lips siphoning the banana from between my fingers. "Oh, wow," was all I could murmur. I was melting.

Jerry was still pitching leading questions at the handler. "Do you think a chimp would get along with a rather exuberant German shepherd?"

The handler was still fielding them. "Why not?"

The handler did not, of course, know our dog Ahab. "Rather exuberant!" Ha! Would you call a Beethoven sonata rather catchy? Raquel Welch rather cute? Ahab was a berserker dog who considered himself our last line of defense against a uniformly and increasingly hostile world. His barks were as effective as most dogs' bites. No one in our apartment house would come near him. They thought he was vicious, and they hated him. If he hadn't been our dog, I would have too.

"Well, hon," Jerry asked, "what do you think?"

"I think you're crazy," I said.

"That's true. But do you like the chimp?"

"Like him? Of course I like him. But I like a lot of things—a house, a trip to Europe, my sanity . . . "

"Ah, baby, just look at him. How can you resist?"

The truth was that I was finding it increasingly difficult. Every time I tried to hand the chimp back to one of the attendants he cried a plaintive "hoo-hoo" and clung tighter.

"Looks as if he's found a mamma," the handler said. Right on cue.

The whole thing was happening so fast and was so ridiculously incredible that it was difficult to believe it wasn't all a setup. (Surely any moment, the man posing as my husband would peel off his devastatingly clever makeup and say "Smile! You're on 'Candid Camera'!") Jerry kept saying things like "Forget about the money, what's important is whether we want to take him or leave him for cancer research, the moon, sadistic zoo keepers, perverted organ grinders." (Was this the same man I had married?) My neck itched. My mind fainted. I needed time to revive my senses, to think, away from the squeak of doggie toys and the scent of exotic animals. Jerry agreed. He pried the chimp from me and put him back in the box. We assured the attendants we would return, hustled Roger and Shep away from the feed barrels where they were surreptitiously waging a monkey-food war, and headed for the nearest cup of coffee. We hoped the caffeine would rally our better judgment, whatever that was.

A half-hour and four cups later, not only had our better judgment failed to rally, it had ceased to exist. We had spent the entire time rationalizing the irrational, and our rationalizations were dazzling. What, for instance, was a mere thousand dollars when we could easily earn that much and more by simply making our pet available for modeling jobs? Not that we wanted to make a model out of him, but if it were economically essential, why not? In fact, we felt so confident that an infant chimpanzee would be in theatrical demand that we even toyed with thoughts of how we would spend his residuals. And what trouble could that little mush-faced fluff ball be, really? All we needed were some disposable diapers and there'd be no cage to clean. Right? A few toys and he'd be happy. Right? Some fruits and vegetables and he'd be fed. Right? Did we have a lot to learn!

Walking back to Trefflich's, Jerry asked Shep if he would rather have a baby brother or a chimpanzee. Most children would opt for the chimpanzee, but Shep was not your usual animal-loving, pet-hungry child. For example, at the zoo he actually preferred the mechanical structures of the cages to the animals. But though chimpanzees weren't much to cheer about, baby brothers were ridiculous.

"Are you kidding?" he asked.

Jerry shook his head. "No. Think about it."

Shep pulled a pellet of monkey food from his pocket and flicked it at Roger. "A Chimpanzee, of course."

"I guess that's it," Jerry said. And it was.

When we returned to the shop, a tall man and a well-dressed woman were enthusiastically cooing over "our" chimp. Something primordial inside me panicked. Loath as I am to confess it, I think it was the same something that goes wild at Lord & Taylor's when someone picks up the dress I'm undecided about. The very thought that we might lose the chimp we shouldn't be buying in the first place cemented our resolve. He was ours!

Jerry calmly edged in front of the tall man and wrote out a hundred-dollar check for a deposit, nodding as the attendant explained that the balance would have to be paid in cash and that chimps were illegal in the city. I petted our baby-to-be and listened to Heinrich's trainer carefully explain how the chimp could wear diapers until he learned to use the toilet. His attributes were endless. Could we bear the wait till Monday to take him home? Not that I gave one thought to what we would do with him when he got there; after all, when you fall in love, who cares what happens next? We taxied back uptown suspended in a weird state of quasi reality marred only by the squeaking of a toy that Shep had bought for Ahab and the crackle of Roger's bag of hamster food.

* *Spreading the Word* *

Back in our apartment, we immediately began planning for our new arrival. He needed a name and a place to stay. The names ranged from Zerbino to Boris. The places ranged from converting our linen closet into a cage to making a corner of our dining room into ape country. We finally settled on *Boris* as a name and on entertaining dinner guests less frequently. It was time to telephone our loved ones and friends to tell them the good news.

Our first call was to Jerry's folks. We couldn't decide whether or not to spring our fantastic news by opening the conversation with "Guess what?" (which unfailingly would elicit a nervous "You're pregnant!") or simply slip it in as a "by the way." We agreed on the latter. After a few minutes of friendly family chatter, we dropped our bombshell. The effect was awesome. My mother-in-law careened out of control as she delivered a scathing indictment of apes in general and chimps in particular, grouping them all under the handy heading of dirty, disgusting monkeys. Her tirade climaxed in an impassioned desire for him to fall into the toilet and drown and in a threat to cancel the order for a dishwasher that was to be our anniversary present. (Was a chimpanzee worth a lifetime of dishpan hands, I wondered?) Needless to say, the conversation was downhill all the way after that. So much for loved ones.

> Madame Abreau . . . undoubtedly the most outstanding great ape enthusiast in recent history . . . was subjected to a great deal of ridicule not only from her fellow countrymen in general but also from her best friends.
>
> *Men and Apes*

As we continued our phonathon for Boris we were surprised and bewildered, not to mention hurt, by the fact that some of our best friends shared my mother-in-law's feelings. We recognized in the style and manner of their responses a hostility that seemed to be based on the very same anthropomorphic qualities that we loved in chimps, a hostility born of a powerful desire to deny our animal heritage.

At rough count, four out of every six callees used the words *spit, bite, dirty, crazy,* and *ichh!* in sentiments that varied from condescending amusement to unqualified disgust. Several felt impelled to regale us with horrendously infantile sex jokes. All too many wondered how we could put diapers on over his tail. Everyone seemed to have at least one anecdote of chimp violence that they simply had to impart. The stories swung from one about a friend of a friend who had had her nose bitten off to one of a child of a friend's friend who had lost the use of her right thumb to another of a friend of a child's friend who had contracted some rare disease, and so on. The key "fact" was that chimpanzees were incurable sexual exhibitionists. Putting all the information together, we came up with an image of a spitting, biting, dirty beast that chomped noses, ate thumbs, and gave kids rare diseases while soiling living room rugs and whacking off! Our little fluff ball, Boris?

* *Ex-Post Facting* *

It was time to find out just how far that trip to Trefflich's had taken us. We wanted an ape gospel but settled for an old *National Geographic* containing part of Baroness Jane van Lawick-Goodall's famous study of chimpanzees in the wild. Despite the fact that I had mentally accepted old trained-and-toothless Heinrich to be a chimp, the sight of that pretty baroness amid a tribe of enormous and frightening creatures that she insisted upon identifying as chimpanzees chilled my brain. Her description of "charging displays," a little something that chimps do when they are socially excited or frustrated, sent shivers down the length of our apartment. "They rush about wildly, hurling rocks or sticks, dragging branches, slapping the ground or stamping, and leaping into the trees to sway branches violently from side to side" was the way she put it. Now substitute books and ashtrays for rocks and sticks, potted plants for branches, pole lamps for trees, and you've got—let's face it—very casual living quarters, to say the least. But I didn't scream until a few paragraphs later.

"Sixty bananas at one sitting!"

Jerry patted my knee. "That's in the wild, and she's talking about a full-grown chimp."

"Of course it's a full-grown chimp. It would be a full-grown anything, eating sixty bananas at each meal! A sea horse could become Godzilla on sixty bananas a meal!"

"You're over-reacting," Jerry said.

We needed more facts. The first thing I did the next morning, after brushing my teeth, walking the dog, and coping with the realization that the dream I thought I'd had about buying an ape was not a dream at all, was to go to the Museum of Natural History.

There was a bastion of anthropological truth, only a few blocks away and open on Sunday. I came home with every book they had on apes and any book that had *chimpanzee* in the index. They gave us the facts all right, straight between the missing link. The more we read, the more we realized that we had made . . . well, probably the biggest goof in the annals of man.

Cathy Hayes and her husband, a psychologist at the Yerkes Primate Laboratories in Florida, had reared a baby chimpanzee in their home. The book she wrote about their experiment with Viki, which included teaching the ape to talk, was enough to discourage any rational person from following in their footsteps.

> In the very beginning, our windows were bordered by drapes, which hung clear to the floor in graceful folds. But at 4 months of age, Viki began using them to pull herself erect. . . . I snipped the drapes to windowsill length . . . then she began pulling herself up to the windowsills. . . . The final lopping off occurred months later, when she entered her Social Phase. Performing for visitors in the front window, she would often swing Tarzan fashion. . . . At that point, I reduced them to a mere valance.

The havoc and wanton destruction that could be wreaked upon our apartment was nothing compared to the physical and emotional strain we were assured. According to Vernon Reynolds, author of *The Apes,* the attachment of a baby ape to a human surrogate parent was so strong that there were cases of insecure baby apes clinging to human legs twenty-four hours per day for a solid month. (Good-bye pantyhose, long walks in the country, bicycle rides, dreams of being a Rockette. . . .) Some foster mothers of apes had to hire constant companions for their babies, and others found that leaving a young one alone for any length of time caused him to lose his will to live. Then there was the little tidbit about home-raised apes often needing to sleep in the beds of their human parents!

That did it. Or if that didn't do it, the combination of horror stories did. It was bye-bye Boris, and probably our hundred dollar deposit, but we both agreed it was certainly cheap at the price.

Monday morning I left for the office, happily anticipating shrieks of delight and amazement when I related the events of our remarkable weekend and secure in the knowledge that Jerry would ease us out of our anthropomorphic commitment. So secure was I that when Jerry called to tell me that Henry Trefflich had no qualms about returning our deposit, I wasn't even suspicious. So secure was I that when Jerry told me he was just going to phone a few people who owned chimps in the city, I wasn't even wary. So secure was I that I barely heard Jerry tell me he was going to visit some ape-owners on Wednesday. On Wednesday night that security all changed.

A piece of white bread was clamped between my teeth and I was deftly chopping onions when Jerry sneaked up behind me and kissed me on the neck. "How's the greatest wife in the whole world tonight?"

"Ummmmm. 'errific," I grunted, keeping the Bond sandwich slice between the onions and my mascara. " 'ow 'uz your day?"

"I saw two girls and a chimp." He kissed me again.

" 'at's nice." I passed the coarsely chopped stage and moved into mincing. (Once I get a rhythm going with onions, nothing stops me.)

"They let me play with him. He was great, friendly, rode a tricycle, too. And they had no trouble with him at all."

" 'at's swell." The old bread-between-the-teeth routine was failing. My eyes began to water. I started to wonder why I was being kissed so much.

"Don't you think it's fantastic? Two girls handling a chimp by themselves? No problems at all?"

"Uh-hum." I scraped the onions into the pot and took the bread out of my mouth. "You know, everyone in my office cracked up when I told them that we almost bought a chimp."

Jerry stopped kissing my neck. I turned around. He just stood there looking enigmatic. Then, without even giving me a chance to think, turn the flame on under the stew, or hide, he said, "What do you mean 'almost'? We pick Boris up tomorrow."

Chapter 2

* *The Homecoming* *

The excitement of bringing a newborn child home from the hospital is nothing compared to the manic frenzy of preparing to pick up a baby ape, especially when the pick-up price is more like a ransom. While Jerry was out convincing a friendly bank teller that he understood all the advantages of a certified check but still preferred the nine hundred dollars in cash (the teller offered small bills, sensing a bit of skullduggery), I raced like a lighted fuse through Gimbel's basement. Baby apes, we'd been told, required as much warmth as baby people, so I shopped accordingly. Blanket, bunting, diaper pins, plastic pants, wool hat, snap-on undershirts ("Excuse me, what size are these?" "How old is your baby?" "Er, I'm not sure." "YOU'RE NOT SURE?"), shirts, a stretchy crawler suit, and disposable diapers. Shaking with excitement, the cold, and the fact that I'd just spent thirty dollars on clothes for a chimpanzee, I hailed a cab and headed for the pet store.

Everyone in the shop was grinning when I entered. Naturally; Jerry had already popped the nine biggies to them. Boris, on the other hand, looked terrified. His hair stood out in all directions and

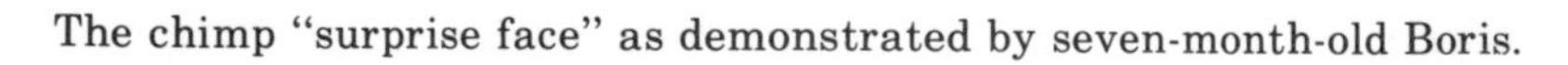

The chimp "surprise face" as demonstrated by seven-month-old Boris.

his lips were pursed in a frightened pout. I picked him up and snuggled him. His face felt wet.

"Why is his nose running?"

"Don't worry about that," said one of the handlers. "We gave him a B-12 shot, just keep him outta drafts. He'll be fine."

"Are you going to be fine, little fella?" I tickled his white fuzzy chin. He sneezed.

"He'll be swell," the handler assured us again. We believed him.

I diapered and dressed Boris, who seemed to enjoy the whole process, slid him into a bunting, and put a woolen hat on his head. He looked grotesque. (It was the hat that did it.) But he was ours, and out into the snow we went to hail a cab and face the future.

On the way to the apartment we hastily devised a plan for introducing Ahab to Boris and vice-versa. "When we get to the door," Jerry said, his voice dropping to a Bogart tone, "you hold Boris while I put Ahab's leash on. Then you bring the chimp into the apartment. *Slowly.*"

Which is what I did. But the moment Boris saw Ahab, he let out a "Hooo," which instantly canceled our dog-sniffs-and-makes-friends plan. All Ahab could see was what looked to him like the world's niftiest squeak toy. And he wanted it! A lot! It was not easy convincing him that this was his new baby brother.

Jerry held the leash with both hands. "It's all right, Ahab! All right, boy."

It was obvious that Ahab did not believe him. That low, throaty growl indigenous to German shepherds and grizzly bears persisted. Boris bristled like a porcupine and tightened his grip on my neck. I began to step backward toward the door.

"Don't retreat!" Jerry yelled. "Ahab will think he's won."

"*Think?* Let's change places," I suggested.

"Just lower Boris and I'll let Ahab sniff him."

"Ahab doesn't want to sniff him. He wants to eat him."

"I'm holding his leash."

"I'm holding his dinner," I wailed.

"You're conveying your anxiety. Dogs sense that. It'll make things worse. Now bend down and—it's all right, Ahab—I'll inch forward."

An adult male German shepherd can throw 750 pounds of pressure into his bite, rip through muscles and tendons and, if he strikes at the right angle, break a collarbone or crush a forearm. For the record, he can kill a man in less than 30 seconds if he attacks the vital organs in the stomach area.

The Guard Dog

Jerry arched backward as Ahab strained toward us. I bent down and let Boris into scent range. Ahab sniffed, sniffed again. Jerry relaxed his grip on the leash. I relaxed my grip on Boris. Boris was taking no chances, and kept a one-arm constrictor hold on my neck.

Jerry unhooked the leash and Ahab continued his olfactory exploration. "There, you see, they'll get along fine."

"I see," I said. But no sooner had I said it when Boris brought his tiny hand down *whomp* on Ahab's snout.

Ahab barked. Boris "hoo-hoo"ed. Jerry yelled "No!" and I closed my eyes. When I opened them, Ahab gave me a look of bewildered disdain and trotted off to find his chewing-bone. The prime advantage in being a pessimist is that things usually work out much better than you expect.

* *Living Quarters, Nickels, and Dollars* *

Most eight-year-old boys have bedrooms filled with adventure books, toys, and gerbils. Shep's bedroom looked like the back of a TV repair shop. Batteries, wires, and transistors were in frightening disarray on his shelves. When we entered, he was kneeling frog-fashion on the floor, where several metal boxes were making ominous clicking noises.

"Oh, you brought the chimp home." He picked up a screwdriver and began to activate another metal box. "I'm making a computer."

"Shep, Boris is going to sleep in your room until we get a cage for him," I said. "You'll sleep in the living room."

"Okay with me, but he better not touch any wires or he'll get electrocuted, and you'll have to buy me new wires."

Jerry placed the carrying cage on top of Shep's bureau. The cage was made of plywood with four air holes on each side and a small screen door in front. We shredded newspapers and stuffed them inside.

"Why don't we give him a blanket instead?" I asked.

"Newspaper is warmer." Jerry tore another sheet. "He can burrow into it. It's like a bed of leaves."

I failed to comprehend the virtues of a leaf bed, especially one made out of the Sunday *Times*, but I deferred to Jerry's knowledge of such anthropological mysteries. We prepared a bowl of lettuce, raisins, and peeled apple slices and watched Boris pick, choose, and finally devour it all with excited "uh-uh-uh-uh" food barks. (A chimpanzee makes a very distinctive noise at the sight of food, which has been described as sounding something like the grunt of a young pig but is actually closer in tone to the guttural cough of a '55 Chevy on a cold morning.) I mixed the formula for his bottle and when he saw me holding it he raised his arms and began an even more excited volley of "un-uh-uh"s. I laid him in my lap. With both his hands and feet clutching the bottle, he drank it dry. I was thrilled (what mother wouldn't be?), but still upset about his runny nose. I made Jerry promise to take him for a checkup the following day.

We searched Shep's bedroom for drafts, found it secure, turned on a vaporizer, and put Boris into his cage.

"You know," Shep said, "he ought to pay rent for using my room."

The little bandit. "Rent?"

"Sure. I could be inventing something that's worth a lot of money if he weren't in my room."

"You can still use your room. You're just *sleeping* in the living room," I said.

All Boris's toys were nontoxic—but not one of them was chimp-proof.

"Some of my best ideas come at night."

We were five cents a night guilty enough to pay the rent. Undoubtedly, the world would miss out on a great electronic invention, but so it goes.

The next morning I fed and diapered Boris. He was playful, happy, and went scrambling in a hunkering crouch all over the bed. Often he would bend forward and trace the stark white lines on the bedspread with his finger. I picked him up and hugged him and he hugged back. It was delicious. I had bought a short plastic pole with rings on it (the toy manufacturer said it was for ages six months to three years), and Boris loved it. He immediately removed the rings and began chewing on them. I showed him how to replace them on the pole and he did it! I was proud, ecstatic, and very late for work. I wiped his nose and put him back in his carrying cage, leaving Jerry to explain to Thelma, our housekeeper, about the newest member of the family.

Thelma Williams was one of those remarkably competent, accepting women who just took things as they were. I'm sure if we decided to flood our apartment and freeze it over, she would simply appear for work with ice skates. Taking care of a wild ape, though, was a lot to spring on anyone. But Thelma was Thelma, and Boris was adorable, and even though Jerry told her that she was in no way responsible for the chimp, he was hers right from the start. He was a seven-month-old infant. She didn't care what kind. An infant was an infant, be it a chimp, a child, or an aardvark. When it was hungry, you fed it. When it was wet, you changed its diaper. And no matter what happened, you never stopped loving it.

I called Jerry that afternoon to find out how things were going.

"Oh, hello, it's you," he said.

Something told Sheena that all was not serene in the domain of the Jungle Queen. "What's up?"

"Ahab's been urinating all over the apartment," he said casually. "He's hostile and upset about Boris."

"I'm hostile and upset about my carpet."

"We'll get a new carpet," he said, and didn't mean it.

"Maybe we'll get a new dog," I said, and he knew I didn't mean it. "How did it go at the vet?"

"Well, it was interesting." He proceeded to explain how he had taken Boris downtown in a taxi and nearly caused a United Parcel truck to deliver itself into a lamppost when the driver saw Boris's head poking out of the bunting. When Jerry arrived at the veterinarian's office, he was met by a slender, nervous man in a white coat who greeted him with four minutes of nonstop "hoo-hoo-hoo-hoo" chimp barks. ("Makes the little feller feel at home," the doctor explained.) The doctor, Jerry felt confident, knew his animals, but just didn't want to show it. The examination was cursory ("Nice chimp you've got there"). The diagnosis was astute ("He has a cold"). The prescription was, assuredly, unusual ("Give 'im Yoo-Hoo chocolate drink. My wife gives it to me. I love it. They love it. He'll love it. Ten dollars, please"). He handed Jerry a bottle of nose drops as an afterthought.

"How's Boris doing?" I asked.

"His nose is still running."

"What about the drops?"

"What about them?" Jerry sounded defensive.

"Have you tried putting them in his nose?"

"What nose? His nostrils are almost flush with his face. The drops just drip out. If it was his upper lip that was congested, we'd be in fine shape."

"You don't have to yell at me. I was only asking."

"Have you ever tried giving nose drops to a no-nosed chimp while he's pulling your glasses off with his feet?"

"I don't wear glasses," I said quietly.

That evening Jerry began his quest for a carpenter to build Boris's cage. One of the major problems was convincing the men he called that he was serious. The other was money. Building a nine-foot-high, five-foot-long, four-foot-wide cage in the corner of a Manhattan apartment dining room was going to be expensive. Jerry finally reached one sporting builder who was willing to do the job for a paltry three hundred and fifty dollars.

We needed a home for a playful little baby chimp who would eventually grow into an equally playful five-foot-tall, one-hundred-thirty-pound, strength-of-six-men chimp. I wanted the cage to be aesthetically pleasing and unobtrusive; after all, it was in our dining room. Jerry wanted it indestructible. We settled with our usual family compromise. It was to be indestructible—tubular steel and heavy-gauge chain links, with bolts sunk deep into the wall. Jerry and I fell asleep that night finally believing that Boris was, for better, for worse, in sickness and in health, ours. Tubular steel in the dining room had that ring of permanence.

* *Dark Awakening* *

Ahab was still sulking over the hairy upstart who had usurped his position as house kink when I took him for his morning walk. He growled and lunged at garbagemen and little-old-lady dog lovers alike. When he had to lift his leg, he kicked it up karate-fashion. Shep wasn't in a much better mood. Even though he was beginning to enjoy Boris (he had given him his rubber Snoopy toy, which is more than he'd do for his best friend), he had just about had it with sleeping in the living room. I agreed to raise Boris's "rent" by another nickel a night. Shep agreed that the living room wasn't so bad after all, and left for school. I filled Boris's bottle and quickly prepared a colorful morning salad for him, heavy on the raisins, and hurried into his room.

I opened the door and gasped. My stomach lurched and tightened into a band of broken glass. The radiator had gone off during the night and the room was freezing. Boris was huddled up inside his cage, shivering, his face wet with mucus. I pulled him out, hugged him to me, and began to cry, rocking him back and forth, rubbing his arms and legs. He looked at me and put out a finger to touch the tears streaming down my cheeks. I wailed, "Jerry!"

"Ohmygod, it's freezing in here." Jerry hammered the radiator with his fist.

I nodded and sobbed. "I should have gotten up in the night and checked him."

"Well, stop crying now. You're upsetting him."

"I can't help it," I sniffed. Boris began sucking on a strand of my hair.

"He's hungry." Jerry handed him a raisin. "I'll try to get a doctor while you feed him."

"Didn't that girl with the chimps you met say that she used a real doctor, not a vet?"

"That's right." Jerry headed for the telephone. "I'll call her now and get his name." He returned a few minutes later. "No luck. There's no answer. We'll just have to find our own. I'll see if I can get one who makes house calls."

The first doctor Jerry called was Shep's pediatrician. No, he did not make house calls and no, he did not want an ape in his waiting room. Jerry tried our G. P. He laughed. The next call was to a doctor we knew whom we suspected of getting his medical license through a mail-order catalog. He wasn't sure if he was the recipient of a bizarrely obscene phone call or merely a put-on. He hung up.

"Try Harold," I suggested. Harold was an analyst friend of ours who knew all sorts of off-beat people. He couldn't help us as far as any medical doctors who handled chimps, but he did say that if Boris had any other problems he'd be happy to oblige.

Meanwhile, Boris had cheerfully eaten all of his food and was gleefully rolling around on the bed, chewing on the bottom of his shirt. He looked fine, but neither of us was convinced that he was. Every book on apes we had read stressed their susceptibility to human infections in general and singled out respiratory infections in particular as the greatest killer. There was only one thing to do, what any other mother does when she can't reach a doctor immediately. I reached for my copy of Dr. Spock. Flipping the pages quickly, I found "The Handling of a Child with a Cold," which said, "Aim to keep your child evenly, comfortably warm. . . . If he is up and around the house he should have as much on his legs as on his chest."

This was going to be difficult. Boris was already wearing a double-thick undershirt, a polo shirt, and a sweater. In order to cover his legs comparably we'd need thermal-lined ski pants. Since his legs were only about eleven inches long, we had a problem. And there was no way he would stay under a blanket.

"How about little woolen scarves tied around each of his legs?" I suggested.

Jerry said it was worth a try, but it wasn't. For the four and a half seconds that the scarves stayed around his legs, he looked like a hairy leper. We tried long socks, cut-up leotards, and finally gave up. Jerry went out to buy a heater.

Throughout the day at the office I scolded myself for being so overly concerned about a common cold. But when I came home and saw Boris, I knew something was wrong. His creamy tan face was flushed and he was hot.

"I'm going to take his temperature," I announced. Sheer bravado on my part, since I knew neither how he would respond to a rectal invasion nor what a chimpanzee's normal temperature was supposed to be. Jerry was impressed. He was even more impressed as he watched me deftly fend off the four simian appendages that were each trying to remove the thermometer in singularly ingenious ways. I looked at the thermometer and panicked. "It's 103 degrees!"

"What's it supposed to be?"

"I don't know what it's supposed to be," I whispered. "But I know it's not supposed to be a hundred and three degrees."

Shep, trailing a morass of wires, entered the room. "What's wrong with Boris?"

"He's sick, dear."

"Uh-oh." Shep dropped the wires and rushed out. He returned with his Polaroid camera cocked and ready to shoot.

"What in hell are you doing?" Jerry shouted.

There was a soft click and a flash. "I thought I'd take a picture of him in case he dies." He began counting to ten, picked up his wires, and left.

It was no time to try and convince some doctor that chimps were just like babies and that they were man's closest relative and how could you desert your own relative in time of need. We pushed his arms into two sweaters, wrapped him in a blanket, and put the whole immovable package into a bunting. We were off to the Animal Medical Center. On our way down in the elevator, two ladies asked to see the baby. I pulled the hood of the bunting down over Boris's face and told them that he was sick.

"Couldn't I just take a peek?" the older one asked. "I love babies."

As the elevator door opened, I pulled back the hood and turned Boris toward them. The older woman sputtered. She was trying desperately to expel some sort of polite comment. We were on the street when she finally got it out. "He's . . . he's very nice. Really."

The New York Animal Medical Center is one of the world's finest animal hospitals. Aside from its pleasant décor and an overall sanitary condition that puts most people hospitals to shame, the patients are given round-the-clock attention and the best medical treatment available. It is open twenty-four hours a day for emergencies and is ready for anything. We arrived after regular clinic hours, and the waiting room—clean, green, friendly—had only a few sick dogs and troubled owners in view. It was very quiet.

We sat down and carefully unwrapped Boris. His soft, silly ears were a bright pink and his eyes looked like brown marbles in a well. He played lazily with my fingers, bending and tasting them. A poodle with a bandaged paw came over to sniff and Boris let out a soft "hooo." That did it; people appeared from nowhere—cleaning men, interns, clerks on their dinner breaks were all around us.

"Is he a monkey?"

"No. He's an ape."

"Is he sick?"

"Very."

"He doesn't look sick."

"Sue me."

"What does he eat?"

"Human flesh!" Jerry snapped, and opened the third pack of cigarettes that evening.

We were finally ushered into an examining room. Boris sat quietly on the table as the doctor put a stethoscope to his chest. He listened for a long time. Too long. At last he lifted his head and the stethoscope hung limp around his neck. "You have a very sick chimp."

The broken glass was back in my stomach. "He had a cold when we bought him."

"He has pneumonia now. When did you buy him?"

"Four days ago."

"How much does an animal like this cost?"

Jerry put his arm around me. I put my arm around Boris. "A thousand dollars," I said.

Boris worked the buttons on the doctor's coat, mouthing them occasionally. "Will he be all right?" Jerry asked.

The doctor was young. His face was not professional enough to mask bad news. "Did you get him with an insurance policy?"

I started to cry and couldn't stop. I excused myself from the examining room and stepped out into the waiting area. A heavy-set woman with a chihuahua stood a few feet from the doorway. "Look, Snoopy. Look at the monkey." She pointed the dog's muzzle at Boris as if she were drawing a bead on him. "Is your monkey sick?" I nodded and began sobbing again. "You really should get hold of yourself," she said. "I've had dogs put to sleep. I know. Why don't you try telling yourself he's only a monkey. . . . "

"He's not a monkey, he's an ape and he's my son!" I shouted and rushed, trailing tears, back to Jerry's arms.

A pretty, round-faced girl in a white nurse's coat was holding Boris. She was smiling as he hesitantly examined her nose with his forefinger. "He's a beautiful little chimp," she said. "We'll give him lots of loving, don't worry."

I forced a smile. Jerry thanked the doctors and said "Good-bye" and "We'll be back" to Boris. He stroked Boris's hand for a moment, then we left.

Jerry didn't cry until he read the phone message that the contractor was coming the next day to start work on Boris's cage.

Chapter 3

* *In the Province of Positive Thinking* *

Although the report the following day from the Animal Medical Center was not overly encouraging, we resolved to proceed positively. It would have been difficult not to. At four-thirty that afternoon Louis Tourget, the carpenter, arrived, and suddenly there were nine-foot-long tubular steel frames stretched across our living room floor, along with enough sixteen-gauge chain link fencing to safely incarcerate a rutting grizzly bear.

Mr. Tourget was a stocky, jovial French Canadian who regarded his craft with the keen zeal of a dedicated scientist. The logistics of designing a cage for a great ape in a Manhattan apartment excited him the way microbes must have enthralled Louis Pasteur. Jerry had stressed that we wanted something roomy, solid, and unobtrusive, and Mr. Tourget had been enticed by the challenge. It was definitely a challenge. The roomy, solid, unobtrusive cage was going to take up almost a quarter of our dining room.

"Your monkey, he will love it," Mr. Tourget assured us.

"Ape," Jerry corrected.

The distinction went unnoticed. "I am going to build a cage that will keep him happy for years."

"It will have to be really strong. Chimps get pretty vigorous sometimes."

> Even when everything had been sunk into cement, a few of the adult males still managed to break something. J. B. was the worst offender. He kept breaking off the steel handles of the levers, so we could not close them. And he managed to snap even strong cables, though the only part showing was a length of about seven inches between the cemented end of the pipe and where it was attached to the lever. It was a rather terrifying indication of the tremendous strength of these chimpanzees.
>
> *In the Shadow of Man*

"Oh, it will be strong all right. When I am through, you can put six Green Bay Packers inside and it will take them more than a year to get out. Your little fellow's tail will collapse before my cage."

"He doesn't have a tail," I said gently.

"Oh. What a shame. I would not worry, though. My brother-in-law, he lost two fingers on a cutting press and he still plays the guitar. You would not happen to have a beer?"

Jerry brought one from the refrigerator. Mr. Tourget flipped off the top of the can with telling familiarity.

"What Hester meant," Jerry said, "is that chimps don't have tails."

Mr. Tourget drained the can, plopped it on the table. "Well, that is nature for you, isn't it? Crazy." He picked up his drill and headed for our dining room.

Ahab intercepted him with a chest-rumbling growl that curled my spinal column. Before we could shout a warning, Mr. Tourget patted the dog heftily on the rump. "Good pup," he said, and to our amazement—and Ahab's—he whistled his way blithely to his work.

Since Mr. Tourget could spare us only a few hours in the evenings, the building of Boris's cage took five nights and as many cases of beer. As the reports from the Animal Medical Center grew more encouraging, we began getting really enthusiastic about it. By the night it was completed, Boris was well out of danger and we were convinced that a nine-foot-high, five-foot-long, four-foot-wide cage was what we'd always wanted in our dining room. And we decided to celebrate.

I was two and a half Scotches into the celebration, which Jerry and Mr. Tourget had begun at least an hour before, when Ahab began to bark and hurl himself frenziedly against our apartment door.

"That's odd, I didn't hear the bell," Jerry said.

It wasn't odd at all. One rarely heard a bell with Ahab around. His bark, which obliterated sound, usually began when he heard footsteps in the lobby, six floors below. Only the most stalwart muggers worked our building.

"Will somebody answer the door? I'm trying to watch television," Shep shouted from the bedroom.

"Were you expecting anyone?" Jerry asked.

"Uh-oh. I forgot."

"Uh-oh you forgot what?"

"It's Milt. He called earlier and said he was coming over to talk to you about our insurance policies."

"Here? Milt Tapper?"

Mr. Tourget broke into a bawdy Canadian logging song. Ahab was doing the chorus. I wished I was somewhere else. Milt Tapper

Boris's cage completed—and our once-elegant dining room finished.

was our insurance broker, a kind, soft-spoken family man from Long Island, born honest, the sort whose idea of casual was to unbutton his suit jacket. He was the kind of man who, because he was committed to conversing about death and making its fringe benefits appealing, went overboard about accepting as routine any life-style he happened into. We were about to put his Penn Mutual cool to the test.

Jerry put Ahab into a "down-stay" about three yards back from the door.

"Hi, Milt," Jerry said easily.

Ahab's growl sounded like a tank gun swiveling into position.

"Oh, you have a dog," Milt said. A perfectly normal understatement. He started forward.

So did Ahab.

"Down, Ahab!" Jerry put his hand on Milt's arm as Ahab reluctantly returned to his quivering sphinx position. "Just walk straight ahead to the living room, he won't bother you. Don't try to pet him, he's wary of strangers. Ignore him."

Ignoring Ahab in our hall was about as easy as ignoring an anaconda in your bathtub, but Milt was amenable to the suggestion and walked without sudden moves to the safety of our carpeted living room. (We had miraculously trained Ahab with the command of "Hall!" to immediately vacate any carpeted part of the apartment. We trained him with inducements of large, juicy soup bones that he refused to part with and so often formed a chilling ceremonial pile in a corner.)

Jerry interrupted Mr. Tourget's song to introduce him to Milt, who was staring curiously at the focal point of our celebration.

"What's that?" Milt asked.

"A cage for our ape," Jerry told him.

Milt laughed insecurely. You knew he hoped he was being put on and feared that he wasn't. "An ape?"

"A chimpanzee," I said. It sounded less threatening.

"H-how big is he?"

"Small," Jerry said.

"Tiny," I assured him.

Milt appeared less than convinced. He kept eyeing the cage as he removed papers from his attaché case.

Mr. Tourget gathered up his tools. "Solid!" he said, swinging a lug wrench against the metal. Milt stiffened uncharacteristically and busied himself sorting forms as Jerry, humming "Frère Jacques," helped Mr. Tourget. I tried to relax Milt by offhandedly beating him to difficult words like *beneficiary* and *deceased* in my questions about life insurance. Although he unbuttoned his jacket, accepted a cup of coffee, and computed in a small black notebook dazzling ways Jerry could die with confidence, his eyes never strayed far from the nine-foot cage. When Mr. Tourget left, Milt withdrew another set of forms from his attaché case.

"I think," he said, casting a guarded look at Ahab and then back at the cage," you should seriously consider doubling your liability insurance."

* *Out Again, In Again* *

We were euphoric when the Animal Medical Center told us Boris could come home, and Jerry and I raced crosstown with bottle, bunting, blanket, and banana to get him immediately. Boris hooted happily when he saw us, though I'm sure the bottle of formula and the banana were at least partly responsible. As I dressed and cuddled him, the doctor told us they'd estimated his age at between seven and nine months. He also explained that what we'd thought was a cyst over Boris's right eye was in fact buckshot. X-rays had also revealed pieces of buckshot imbedded in his back. As we recalled that apes spend most of their first three years clinging to their mothers, the circumstances of Boris's capture became tragically clear. It cemented our resolve to make reparation as best we could to this soft wild innocent for the birthright he'd been denied.

When we arrived home, we couldn't wait to show Boris his new cage. It was the most impressive structure I'd ever seen in an apartment, especially with the oak-stained plywood Jerry had put over the two walls that had once been the corner of our dining room and were now half of the cage (so that Boris couldn't damage the plaster) and the plush red rug that covered the floor. But Boris was more than impressed, he was overwhelmed. And small wonder! The cage dwarfed him. When we placed him inside, he looked like a child's toy at the bottom of a pretty elevator shaft; he looked lost. Jerry and I experienced a swift moment of panic. We'd wanted Boris to have plenty of room, the best of everything, but had we given him too much too soon?

Our fears were swept away seconds later in a jingle of noise as

The proud parents home from the hospital with their baby.

Boris bounded up the chain webbing and began happily pounding his feet against it, "hoo-hoo"ing with an exuberance to fill any parent's heart with joy.

For the next few days, the cage became the focus of all household energies: Boris practiced somersaults in it, Jerry designed shelves and poles for it, Shep tried to wire it with an escape alarm, and Ahab urinated on it. I converted a plastic baby bathtub into a snug bed, and Boris's creature comforts were complete.

Once happiness was assured, we began to concentrate on health. The Animal Medical Center had cautioned us about drafts, and we weren't taking any chances: we decided to seal the windows. This wasn't a particularly difficult feat, but it did take its toll aesthetically. Our dining room windows, limned with snakes of gray caulk, looked as if a horde of gum chewers had performed an act of ritual vandalism.

But though the room was now draft free, we couldn't be sure that the building heat would remain consistent, so Jerry bought a thermostatically controlled electric heater and a cold steam vaporizer. After four days, we were all set and everything was running smoothly. Unfortunately, so was Boris's nose.

On the evening of the fifth day, Boris's temperature was 102 and climbing. By 2:00 A.M. it had reached 104, and his caramel-colored cheeks were flushed a raspberry pink. We called the Animal Medical Center and the emergency doctor agreed that we should not wait until morning. While I cuddled Boris on my shoulder and helplessly paced the floor, Jerry ran downstairs to forage for a taxi. He returned twenty long minutes later and reported that a somewhat stoned cabbie, who enjoyed saying "ding-dong" at every meter click, was waiting for us outside. We bundled Boris into his bunting and prayed that we weren't too late as we ding-donged our way across town.

The doctor on duty was encouraging, though the diagnosis was not—Boris had another nasty case of pneumonia. It wasn't going to be easy to whip, but the doctor guaranteed they'd do everything in their power to try. Boris, he told us, had become the darling of the Medical Center.

They drew fluid from his lungs and grew a culture, which revealed the presence of a rare pneumonia virus, usually inclined to old men (seven to nine months?) and unresponsive to any of the antibiotics currently in use there. A canvassing of specialists at the Center unearthed information about a new drug being used on the

West Coast. The Center flew it in immediately and their darling responded with gusto.

When we came to pick him up two weeks later, several of his fans on the staff had brought their cameras and wanted us to take pictures of them holding Boris. He obliged, we obliged, and the waiting room of the Animal Medical Center was suddenly transformed into an instant bon voyage party, with kisses and tears and even a ribbon-wrapped banana. As we were leaving, the doctor informed us quietly that Boris still had some fluid remaining in his lungs.

"His body should resorb it," he said.

"What happens if it doesn't?" I asked.

"Well, there'll be some problems, but I think we'll be able to handle them."

I had no doubt about the doctor handling them; it was us I worried about. Our finances had shriveled thinner than a swizzle stick, and there were already IOUs in Shep's piggy bank. Although neither Jerry nor I wanted to admit it to each other, the darling of the Animal Medical Center was costing us a bundle.

Boris was delighted to be home again. He hooted a friendly hello to Ahab and ignored the slight when Ahab merely growled resentful acknowledgment and skulked off to his pile of bones. Boo was too happy rediscovering the familiar. He touched lamps and pictures tentatively, gobbled down raisins, and bounced joyfully on the couch, sniffing pillows, tasting tables, expressing his pleasure to us in hoots and kisses.

When we brought him to his cage, he took one look at the ropes Jerry had hung inside and dashed to them. With a tiny bound he grasped hold with one hand and began swinging.

"Shep!" I called. "Come and look. Boris is swinging."

Shep, sporting a pair of giant earphones, entered the dining room. He watched Boris for a moment, nodded, then started to leave.

"Aren't you excited?" I asked.

Furniture was for tasting.

"About what?"
"His swinging."
"All apes swing."
"But this is his first time."
"It won't be his last," he said confidently, and returned to whatever he'd come unplugged from.

Granted, a swinging ape wouldn't turn heads in the jungle, but this was our apartment—our ape! Jerry and I were dazzled.

The next morning I discovered that Boris had grown up quite a bit during his stay at the Medical Center. He was bolder and much more active. As I prepared his breakfast, he jumped from the chair and scampered under the table. I thought it was cute, until I realized he had claimed the territory as his private playground. He swung round the table legs, somersaulted, trumpeted hoots like an elephant with hiccups, and refused to come up to eat. He'd dash out, grab a piece of banana, and return, hanging by one arm while he ate, pushing leftovers into cracks with his toes. Ahab watched the proceedings, with stony canine hostility, from the other side of a wooden baby gate that blocked his entrance into the kitchen. Boris waved food and taunted him unmercifully. I told myself it was just a phase and wondered whom I was kidding.

* No Business like Show Business *

> Television is best suited to chimpanzee talent, and this is certainly the field in which they have really made their mark.
>
> *Men and Apes*

Boris had only been home a week when we received a phone call from a Mr. Kahn, supervisor of an experimental TV film group at New York University. He'd gotten our name from the Animal Medical Center, which he'd called in desperation, trying to locate an available chimpanzee. His group was shooting a documentary on anthropoids, to be used in schools as a visual aid, and the professional chimp they thought they could get had decided at the last moment to hold out for more money. They were offering forty dollars.

"I'd really like to help you out," I said, "but our chimp isn't a professional."

"He has to start somewhere," the young man at the other end said.

"You don't understand. He's a baby."
"He's a chimp, isn't he?"
"But he doesn't do any tricks or anything."
"We'll work around it."
"He's just gotten over pneumonia."
"Glad to hear it. Can he get over here on Sunday?"
"But—"
"Look, you're our last hope. We've tried everywhere. We're desperate."
"Well . . . "
"Fine! See you Sunday." He rattled off the address and hung up.
"What was that all about?" Jerry asked.
"Boris," I said. "He's going to be a star."

A few trained sub-human primates have reached the top of the entertainment business, because of their exceptional talents or exceptional managers.

Men and Apes

I knew as I headed downtown that Sunday that I didn't have the makings of a stage mother. I did, though, have the makings for an easy dozen banana milk shakes and probably as many health salads. I'd crammed a flight bag full of enough bananas, grapes, apples, lettuce, Yoo-Hoo, and formula, along with Pampers, sweaters, and socks, to take Boris around the world, but the address I'd been given was only a few blocks below Fourteenth Street.

Boris enjoyed the ride as much as did the cab driver, who, I was certain, drove at least sixty blocks looking backward. He persisted in trying to shake hands with Boris, which is an apparently innate human response to meeting apes, monkeys, friendly dogs, and occasionally seals. Boris persisted in trying to eat the cigarette butts in the ashtray. I was frazzled when we arrived at the studio.

Mr. Kahn met us at the entrance. He was a slim man in his early twenties with incipient élan and a budding professional smile. He extended his hand to Boris with the conscious earnestness of an adult greeting a lad back from Eton. Boris's response was less than Etonian. As Mr. Kahn led us up to the studio anteroom, Boris gripped my neck tightly and sucked his forefinger. (Most infant chimps, by the way, suck their thumbs, whereas infant gorillas suck their forefingers. Boris always sucked his forefinger. It comforted me to think that if he was insecure, it was only because he felt he should have been a gorilla.)

At least when Boris felt insecure, he was macho.

Photo: David Sagarin

The taping of the first half of the program was still in progress, but we were told it wouldn't take long. I hoped not. The studio's air conditioners had been set for the comfort of menopausal polar bears. Boris was safely bundled in his wool bunting, but as the minutes ticked by he grew increasingly anxious to shed it. He was intrigued by the new surroundings and entranced by the attention he was getting from the cast and crew. Bob, the producer of the show, was Boris's favorite. He was a soft-spoken young man with blond hair. Though Boris shied back from other males who tried to hold him (never females, that canny imp), he was happy to go to Bob. The producer assumed it was his charm that captivated Boris, as did I. Only later did I discover that Boris had a definite "thing" for blonds. He scrutinized Bob's hair, lifting strands and letting them fall through his fingers, like a shampoo consultant at Kenneth's.

Due to a conspiracy of production snares, Boris's big moment was a long time in coming. I attempted to assure myself the wait was worth it, but after two and a half hours of trying to keep Boris reasonably inactive by pacifying him with food and an assortment of miscellaneous items from my purse, I was unconvinced. However, the moment finally came, and Boris was in remarkably high spirits, shaking my key ring with the fervor of a lead maraca player as we followed Bob into the shooting studio, which to my dismay was even colder than the anteroom. Though the whole point of Boris's being there was to show an ape in the flesh, I didn't dare remove his sweaters. This caused much consternation and a long huddle between Mr. Kahn and Bob and Mr. Kahn's wife, who was to be the moderator. When they broke, they agreed there was no reason to endanger the animal's health. The students who would view the film would just have to take it on faith that hunched beneath that oversized wool sweater was the same animal diagramed in the poster hanging in the background. Besides, they could always go to the zoo.

Bob draped a microphone around my neck and explained that Mrs. Kahn would simply ask me a few questions about Boris, who they hoped would do a few elementary ape things during the interview. Boris did a premature impression of a chimp snake swallower by trying to eat the microphone. I removed it from his mouth and slipped the little ham a grape.

Bob led us to a table and I attempted to position Boris so he'd face the camera. It wasn't easy; Boris wanted the mike. He continued to swivel around, grabbing the wires with hands and feet, until they turned on the spotlights. Then suddenly his interest shifted: he wanted to eat the light beam. He grabbed at it with his hand and hooted softly, confused.

"All right. Are you ready, Mrs. Mundis?" Bob shouted.

I panicked. "You can't call me that," I said suddenly.

"Excuse me?"

There was silence, the whole crew eying me suspiciously.

"I mean, you can't call me Mrs. Mundis." Recalling what Trefflich had told us about wild pets being illegal, and remembering a heart-breaking article in *The New York Times* about a confiscated cougar, I was paranoid. "It's against the law to own a chimp in the city. Some nut watching the show might report me."

"This is a documentary for schoolchildren," Bob said defensively.

Schoolchildren in Nazi Germany had turned in their parents, hadn't they? I wasn't taking any chances. "You'll have to call me by another name."

"All right, Mrs. Jones."

"It sounds phony."

"It is phony. Look, Mrs. Mundis, we—"

"How about Mrs. Roberts?"

"Swell. Are we set now?"

I nodded. Mrs. Kahn nodded. Boris jerked the wire to my neck mike and sent the sound engineer into a controlled fit. Somewhere behind the lights there was a muttered obscenity.

Bob gave the signal and Mrs. Kahn turned to me with a practiced academic smile and began asking questions about Boris's anatomy, his similarity to a human infant, his living quarters, and so forth. I surprised myself by knowing a lot more about chimpanzees than I'd thought, and everything was going well until it came time for Boris to demonstrate the few elementary things he'd been hired for.

One of the most remarkable solo chimpanzee performers of the twentieth century was a six-year-old male called Peter, who appeared at the New York Theatre in 1909. His long routine was purported to have involved no less than fifty-six separate acts, executed in an ex-

> act sequence. Peter entered, bowed to the audience, removed his cap and settled down to an elaborate meal. When he had finished, he smoked a cigar, spat into a cuspidor, cleaned his teeth, brushed his hair, powdered his face and tipped his keeper. He then undressed, lit a candle, got into bed and snuffed out the light. But soon he got up again and put on his trousers and roller skates in order to chase a young woman about the stage. The performance concluded with a fifteen minute display of trick cycling. As a finale he drank out of a tumbler while still pedalling furiously, then hurtled around the stage with much flag-waving before dismounting, clapping his hands, and making his exit.
>
> *Men and Apes*

Every time I tried to make Boris stand, he'd leap up, fling his arms around me and make an oral lunge for the microphone. Mrs. Kahn ad-libbed a lot with nervous laughter. The only thing Boris did on cue was eat a banana—but with style!

When we watched the completed tape, Boris pointed to himself on the monitor but evidenced no excitement. (I feared his reaction was not dissimilar from what the schoolchildren's would be when the program was viewed.) As we were leaving, Bob, whom Boris had smothered with kisses all afternoon, came to say good-bye and handed Boris two crisp twenty-dollar bills. Boris hooted angrily, as if he was insulted, thrusting his face indignantly forward and barking furiously. He refused to touch the money. Bob was hurt. I consoled him by saying Boris was above taking payment for acts of love, and slipped the bills into my pocket.

Leaving the studio, Boris and I were trailed by a group of insistent youngsters clamoring to "pet the monkey." Boris hooted encouragement as I tried to hold them off, hang onto him, and hail a taxi at the same time. Two cabs passed us by, and I can't say I blamed them; we looked like weirdos on the lam. One finally stopped and I collapsed into the back seat, restraining Boris, who was determined to sit up front. The driver was unimpressed. All he wanted to know was whether he should mark us down as two passengers or one. I told him it was a philosophical question that could be argued either way. He wrote down one and was silent for the rest of the trip.

Boris got his second wind before his first had expired, and exploded in a burst of energy when we arrived home, somersaulting, swinging, stomping my headache into a dilly, right up until bedtime. One of us, I feared, as I reached for the aspirin, was not cut out for show biz.

Chapter 4

* *A Joyous Insanity* *

> To make a success of rearing chimpanzees one must devote day and night to them, accede to all their demands and become as far as possible a mother ape.
>
> *My Friend the Chimpanzee*

A new baby in a household always means a few changes in routine. A new chimp, on the other hand, radicalizes any former life-style beyond recognition. Keeping a full-time job and a full-fledged ape was a joyous insanity, and fortunately Jerry and I were just crazy enough to love it.

Mornings with Boris were always an adventure, which is not to say an adventure on which everyone would hunger to embark. My 6:30 A.M. ritual was to dress quietly and tiptoe down the hall past the dining room. If the door was even slightly ajar, I'd have to pull it closed with a hanger so Boris wouldn't see me and start screeching before I took Ahab for his morning walk. Both animals were extremely concerned with priorities. I was concerned with getting Ahab on and off the street before too many moving targets started for work. If Boris didn't see me, he'd lie in his bed and babble low "hoo-hoos" to his toes, but if he did see me, most of the tenants of the building would hear about it in an ear-splitting shriek unparalleled by anything save a short-circuited public address system.

Once I made my appearance, Boris would jump to the chain link, pummel it with his feet, and hoot until I went to the desk drawer where we kept the key (I always felt like a guard in one of those 1940 prison movies). Once the key was in my hand, Boris would push his face close to the lock and watch me intently—very intently. Then, when the cage door was opened, he'd spring out with his arms raised, climb aloft, and greet me with explosive "hoo-hoo"s. And a very soggy Pamper.

Diapering an ape is an art and a challenge. You have to be creative, and speed is essential. After preparing Boris for what was about to happen with a cheery "Diaper time!" I'd lay him down on the couch and give him a good friendly tickle. Then I'd hang onto his two hands with one of mine while battling his feet and removing the diaper with the other. One long sneeze or a short sideways glance and he'd jack-knife backward and be off and running. (I could usually count on Ahab to head him off at the pass.) When I caught him, I'd bring him into the bathroom for a quick cleanup in the sink. Dawdling at this phase could cost me anything from a sixty-nine-cent tube of toothpaste—chewed, squashed, or deposited in the toilet—to a forty-dollar-an-ounce bottle of perfume, drained or shattered on the floor. Of course, there was always the chill factor. Before I'd even reopen the bathroom door I'd have to be sure that Boris was perfectly dry and wrapped securely in a towel. Then it was back again to the couch for a new Pamper and clean clothes. Though Jerry and I—and Boris, I'm sure—would have preferred to keep Boris au naturel, we couldn't. He needed a shirt for warmth, and without overalls he'd pull his Pamper off in a minute—and in our dining room that just wouldn't work.

After Boris was dressed, I'd bring him into the kitchen. While I warmed his bottle and prepared a bowl of lettuce, apples, and bananas, he'd chew on a piece of raisin bread and raid the cabinets for pots and pans. He'd drag them out onto the floor and bang them with a wooden spoon. When he tired of this, he'd bang his head with the spoon, then the table, the floor, my shins. If Ahab made the mistake of coming to the other side of the gate to investigate the noise, Boris would get him too. By the time I arrived at my office, I always felt as if I'd run the four-minute mile.

> Jane Goodall has pointed out that chimpanzees in the wild rely a good deal on the sense of touch for communication.
>
> *The Ape People*

One morning Boris watched me make a telephone call. He seemed fascinated by the whole process, and no sooner had I hung up than he was on the chair and had the receiver in his hand. Having confidence in the indestructibility of anything connected with AT & T, I let him play. The next thing I knew, an angry male voice was demanding to know who was calling. I grabbed the phone and mumbled something about my youngster fooling around,

Manicures were tricky but fun. Boris accepted them as urbanized grooming sessions.

Human methods of communication intrigued Boris.

apologized, and quickly hung up, fearing afterward that I'd sounded suspiciously unconvincing and that the man, suspecting foul play, would try to trace the call. Boris had a proclivity for fertilizing my innermost fears.

Sunday mornings were Boo's favorite. Like most youngsters, he delighted in roughhousing it in bed with daddy, while Jerry, like most daddies, clamped on his temper, grinned and bore it. Our bed was a regal fourposter with struts across the top for a canopy, which fortunately had not yet materialized. Boris viewed it as a Y away from home. Our orthopedic mattress made a fine trampoline, and the posts were great for climbing. At nine months he couldn't overcome the waxed surface of the poles to reach the struts, but all of us knew it was only a matter of time.

In the bed he'd dive under the covers, or lope around with a sheet drawn over his head, which would unfailingly quicken Ahab to a protective barking frenzy. Ahab's protective barks differed from his ordinary ones only in their duration. Convinced that he alone knew the enormity of the danger that threatened, he disregarded commands for quiet with a zealot's conviction. This in turn forced Jerry's "No!" or "Out!" up several decibels beyond acceptable Sunday morning noise levels, prompting outrage from Shep, who couldn't hear his cartoons, and brought a broom-handle staccato from the floor of the apartment above. Boris enjoyed the chaos, tossing a few vocalizations of his own into the fray, though usually from the safety of my arms.

* *Mamma's Boy* *

Boris had clung to me quite a bit when he first returned from the Medical Center. Assuming he was still shaken by his experience, I thought nothing of it. He was insecure, needed reassurance; it would pass. I learned to perform such rudimentary tasks as applying makeup and vacuuming the floor like an amputee. But when more than a month had passed and his need to cling not only failed to diminish but was increasing, I began to worry.

> The normal response of an infant chimpanzee to a sudden movement of its mother is to tighten its grip.
>
> *Primate Ethology*

Every time Boris was out of his cage, he was on me. He straddled my back when I made beds, hung from my neck when I prepared meals, rode my hip when I did laundry, and hugged my legs when I read. If he was playing happily with something and I left the room, he'd drop whatever it was and come "hoo-hoo"ing after me with his arms outstretched. There was no getting away from it, or him. Boris was a mamma's boy.

I decided it was time to cut the apron strings, but Boris had other ideas. I'd soon learn that it was as easy to compromise with an ape as it was to negotiate with a shark.

I'd come home from work one evening feeling particularly hassled. Boris's welcoming din from the dining room began when he heard the faint jingle of my keys way outside the door. (Chimps have a remarkable degree of auditory acuity, an attribute totally inevident when you tell them to do something.) He "hoo-hoo"ed excitedly and rattled the wire of the cage until I let him out. Then, arms extended and nut brown eyes beseeching, he stared up at me.

> The "pout face" is always made by an infant seeking to re-establish physical contact with its mother. . . . In addition the infant may reach one or both arms towards the mother. She normally reacts to this expression and gesture by going to collect her infant.
>
> *Primate Ethology*

Well, I thought, I'll just give him a little hug, change his diaper, and put him down. Well, I thought wrong.

After he was changed and clean, he refused to let me out of his clutches. I tried to coax him into playing with his rubber ball or hammer and pegs while I prepared dinner, but he'd leave me for only a few seconds and then come scurrying back, tugging plaintively at my legs until I picked him up.

Béchamel sauce is tricky enough to get right without the added handicap of stirring it while your neck is full of arms. Something had to be done. Since it was my Béchamel sauce and my neck, I recognized that I had to do it. I gave Jerry a handful of grapes, which were Boris's favorite treat. At this point I felt that bribery was justified.

Jerry opened his arms and held out a grape. Boris eyed him, snatched the grape, and tightened his grip around my neck. I respected the fact Boo was incorruptible but knew now that I'd have to make him an offer he couldn't refuse. Before he had an inkling of what was about to happen, I pried him loose and plomped him in Jerry's lap. He struggled with the outraged fury of the just, ignoring all grapes, all cajoling, ignoring every attention-swaying ruse in the book. And then our little fluff ball got angry.

> The mature chimpanzee Cogo was kept in a compound, but he eventually broke away from it. . . . Once, in a few minutes he wrung the necks of 27 hens, 4 dwarf antelopes, and 2 ground hornbills. In the confined state Cogo was an assassin, but when he was free he was gentle, obliging, and affectionate.
>
> *The Ape People*

Suddenly Boris's hair bristled; his eyes shot black lightning. He stood erect on Jerry's lap, threw his arms in the air, skinned back his lips and rapid-fired raging barks—"uh-uh-uh-uh"—with goose-bumping resonance. Jerry was stunned. I was shaken. Such violent emotion, though we'd read about it, seemed incongruous with so adorable and loving a creature. But incongruous or not, it was there, Boris was capable of it—and to prove it he bit Jerry's arm.

Jerry slapped him. Instantly, the raging ape disappeared and Boris was once again a cuddly, finger-sucking chimp. The mini-rage had passed and Boo seemed to be contrite, but the image of his fury remained indelibly stamped upon us, reminding us in weak moments of romantic anthropomorphism to remember and respect

Boris had no difficulty making himself perfectly clear.

him, despite all his human similarities and infant caprices, for the wild animal that he was and would always be.

With the furor over and all forgiven, I returned to my sauce and Jerry put Boris down on the floor to play. The next thing we knew, a crunch-crunch sound was coming from Boris's mouth, a particularly ominous sound, since I had not yet given Boris his dinner and there was nothing edible in sight. We were to learn that baby chimpanzees were less than gourmets in making distinctions between what was and was not edible. The plastic soap dish that had fallen to the floor was missing a small chimp-sized chunk.

Like a quarterback with six seconds to play, I dove for Boris and hoisted him into my arms. He smirked at me with that inscrutable crescent chimp grin and I knew I was in for trouble. Pushing my finger gently through his lips, I found it could go no farther. His teeth were as clenched as a frigid clam, and he was just about as willing to relinquish his prize. I pinched his cheeks while Jerry tried to force open his jaws, but it was futile. That ten-pound fluff ball had dental clout.

"Maybe if we leave him alone he'll spit it out?" I said, having a tendency toward laissez faire whenever I'm stymied.

"It's worth a try," Jerry said, unconvincingly.

We put Boris down and trailed him around the room, watching him as if a casual sidelong glance would deny us the spectacle of a lifetime.

Boo was deliberately oblivious to us. He loped across the room, knuckles on the floor, using his arms like crutches as he swung his legs forward between them. He pulled open a cabinet door and began banging the lids of saucepans, sounding for a favorite toy. Jerry and I cast each other significant looks. Usually when Boris played, he opened his mouth and hooted. He was now sitting in a large copper-bottomed frying pan, flicking the hook at the end of the handle with his finger. It looked as if he was having fun; his hoot could only be moments away. But Boris's lips were sealed tighter than the windows in our dining room.

Laissez faire, coercion, and divine supplication having failed, I decided to try a banana, a last-ditch attempt I rationalized as a stroke of genius. I grabbed the yellowest one from the table and peeled it halfway down.

"What are you doing?" Jerry asked.

"I'm going to hold it in front of him. When he starts his food barks, you get your finger in his mouth."

Jerry took the banana from my hand.

"Don't you think it will work?"

"It very well might, but I'll hold the banana."

"But he's not used to you as food giver."

"I'm not used to him as flesh taker. Anyhow, you have smaller fingers."

There was no time to argue, and besides, martyrdom had always been one of my more successful sidelines. I knelt down next to Boris.

"Ready?" Jerry asked.

Without even asking for a last cigarette, I nodded.

Jerry flashed the banana.

What ensued was noteworthy only in its predictability. Boris "uh-uh-uh"d excitedly, I got bitten, and both the missing piece of soap dish and the banana went happily and uneventfully down an anthropoidal alimentary canal.

* *Spoonerisms* *

Relieved, but not anxious to encourage Boris in pursuing such eclectic eating habits, I was struck by the thought that teaching him to use a spoon would lead him in safer and more palatable directions. Not knowing the first thing about teaching an ape anything, I was forced to call upon sheer ingenuity: I handed him a spoon.

To my delight and Boris's, he guided it right to his mouth.

"Jerry, did you see that!"

"I saw it."

Boris did it again. Teaching an ape was easier than I thought—I thought.

"Look! He's still doing it!"

"It would be more impressive if there were food on the spoon."

"He's only nine months old," I said defensively.

"I ate by myself at eight months," Jerry said offhandedly, adding "I think," just to cover himself. Jerry was a stickler for accuracy.

"You had a mother to teach you," I said, bravely marching into a losing battle.

"Well, Boris has you."

"It's not the same thing. I'm not an ape."

"No argument there."

Obviously, both Boris and I would have to go further to garner any approbation from Jerry. I plopped a raisin onto the spoon.

Boris took the spoon and slowly brought it to his lips.

"There!" I said. "How about that?" And no sooner were the words out of my mouth than the back end of the spoon was in Boris's, and the raisin was on the floor.

Jerry kissed me. "Maybe it was ten months."

The spoon soon became one of Boris's favorite toys. He loved to suck on it, bang it against his cage, and clutch it with his toes, but eating with it by himself was another matter entirely. It was a matter of how much time I had to clean up the kitchen. Strained plums, which inevitably received Boo's loudest anticipatory food barks, were as much fun for him to dribble on the floor as they were to eat. When allowed to feed by himself, the success ratio was two in the mouth to three on the vinyl. Often he'd jump down from his chair and smear the plums into grand Rorschach designs. Fortunately, I was unimpressed, which alleviated my guilt about inhibiting his creativity whenever I hoisted him back to his seat and wiped up the mess. After several weeks of this, I conceded that I'd attempted self-feeding too soon, and so took to having double-spoon mealtimes: one spoon for Boris to bang with, the other for me to feed with.

Boris had a variety of toys, some store bought, some simply

Every pot lid was a potential hockey puck or Frisbee.

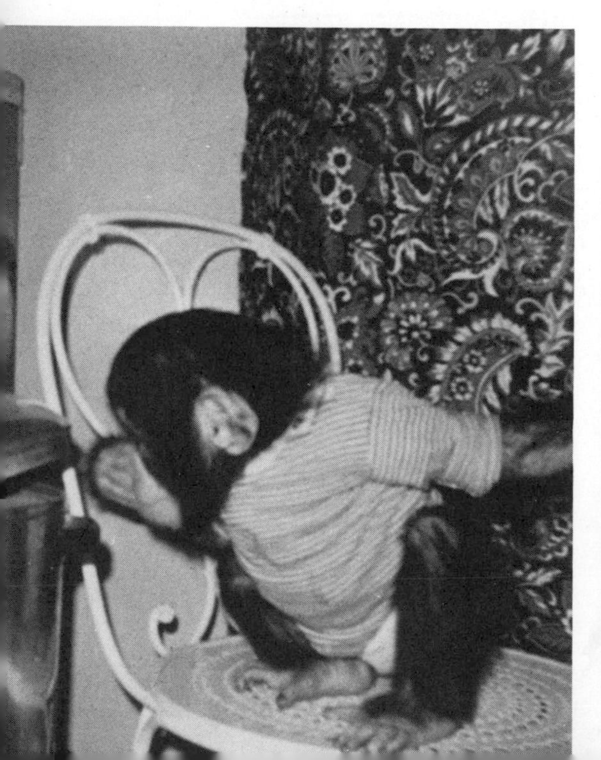

We discouraged Boris's writing aspirations; the typewriter was no match for his heavy-handed prose.

created from what was in reach. Among the store-bought items were "childproof" plastic rings (which were quickly chewed into grotesque pitted shapes that the manufacturer would never have acknowledged as his product), plastic blocks (which met the same fate as the rings), a teddy bear that Boris loved, and a wooden teeter-totter game that he hated. As for other playthings, his supply of amusements was limited only by our no's. Leaves of plants were good for plucking, tasting, and tossing up in the air like paper airplanes. Cigarette packs, despite the surgeon general's warning, were good for chewing; the piano was always fine for noisy jumps, and the typewriter keyboard (the biggest No! of all) was terrific for just monkeying around. Then, aside from pot lids, which made first-class hockey pucks, there were the bookshelves, filled with a seemingly unbounded supply of dropables. Boris heartily enjoyed selecting volumes at random and letting them crash to the floor. And of course there was always our pole lamp for scaling, and Ahab for tormenting. Though we hadn't planned it that way, infant Boris had just about everything. And he wasn't about to share it.

* The Sibling Rivalry Scare *

One evening I put Boris back into the cage and, in order to straighten his bed, picked up his teddy bear. Suddenly, *wham*! he knocked the toy out of my hand. He'd never done anything like that before, so at first I thought he'd suspected I was going to take the teddy bear away. Explaining that I was merely tidying up, I took the bear, patted it, and tucked it lovingly back into Boris's bed.

"See, I like your teddy," I said, and immediately realized that wasn't the problem. Boris picked up the bear and swatted it several times, kicked it, and then hurled it savagely to the floor. Boris was jealous—openly, unequivocally, and violently. And just to make sure he'd made himself perfectly clear, he took the doll in his mouth and disemboweled it.

I returned to Jane Goodall's book that night and found nothing equivalent in the behavior of wild chimp infants. They were either more civilized or their mothers were more intimidating. I took down Dr. Spock and began to read up on sibling rivalry.

A few days after the incident, Shep returned from a long weekend with relatives in New Jersey. He was feeling expansive and anxious to romp with Boris. Since his overtures to Boris were infrequent, I was quite pleased. But as he took the key and went to the cage, I began to feel apprehensive, a feeling magnified by images of Cain and Abel superimposed over a disemboweled teddy bear.

Shep hesitated. "What's the matter with you?"

"Nothing," I shot back, too quickly.

"Why are you looking at me that way?"

"What way?" I asked, in that innocent tone that's a dead giveaway to anyone over the age of six, and a few under.

"Like I'm about to take ten thousand volts from a tesla coil. That way."

"Ridiculous. Look, Boo is waiting for you. Go on, let him out."

Boris's face was pressed against the wire. He thrust his arm out of the cage and beckoned. I had to have faith.

Shep cast me a last distrustful look and opened the cage. I held my breath. Boris jumped into his arms and hugged him, voicing his pleasure with ebullient "hoo-hoo-hoo"s. Teddy bear siblings, at least for the moment, were obviously more threatening than human ones. I let out my breath discreetly.

Brotherly loving. *Photo: David Sagarin*

Chapter 5

* *Day Mommies Are Different* *

My being a working mother had definite advantages for both Boris and me. The most definite advantage was Thelma.

Thelma Williams, our housekeeper, was Boris's "day mommy." She would arrive daily at noon and promptly take her "baby" out of the cage, change and diaper him, wash his hands and face whether he liked it or not, and feed him lunch. Though she knew Boris's regular diet was formula, lettuce, raisin bread, and bananas, she liked to vary it. No one had ever told her that chimpanzees were primarily vegetarians. Even if they had, she probably wouldn't have believed it. We didn't discover until a year later, when Thel happened to mention it, that Boris's lunchtime menu included such unsimian stables as bologna, frankfurters, tunafish salad, chicken soup, toasted cream cheese and jelly sandwiches, and liverwurst! Though Jane Goodall counted it a major discovery to learn that wild chimpanzees actually eat meat, Thelma Williams accepted it as a matter of course.

Boris, like most children and twenty-year army men, liked routine. He looked forward to Thelma's arrival. If twelve o'clock rolled around and Thelma was late or, as it infrequently happened, she could not come to work, he was wildly distressed. Though Jerry would feed him that day, Boo reacted as if his whole world had come undone. He became nervous and frightened, and when I would return from work that evening he would cling to me with skin-pinching fierceness.

One day I came home unexpectedly in the early afternoon. Thelma had Boris on her lap and was tickling him. Boo was laughing short, breathy pant-hoots that sounded like a tiny steam engine or a lover waiting for a go signal. I greeted him with my usual "Hi, Borie-Boru!" and prepared for him to scamper into my arms. I was in for a surprise; not only didn't he come to me, he leaped upon Thelma, grabbed her in a possessive hammerlock, and hooted loudly and angrily at me to go away!

I was hurt. I was convinced he hadn't recognized me. I repeated my greeting, but he only hooted louder, and when I went to take him from Thelma he turned his back and clung more tightly to her.

I had to accept it. Day mommies were day mommies, working mothers were working mothers, and as far as Boris was concerned, the twain were not supposed to meet. When I would return home at five-thirty, he'd leave Thelma and greet me with heart-stirring filial eagerness, but with any arrival before that hour, he'd treat me like dirt. There was no swaying him from these rigid divided loyalties. His routine was as solid as a bronze coconut, and I was about as likely to understand it as I was to crack it.

* *Boris and the Boob Tube* *

Boris knew what his day was supposed to be like, and there was hell to pay for any divergence from this norm. After his breakfast and morning romp through the kitchen cabinets, he happily returned to his cage and waited for me to position the TV in front of it. Chimps

Boris and his "day mommy," Thelma.

in captivity need stimulation to avert apathy. I needed my job to avert poverty. Television was the perfect solution to fill the three-and-a-half-hour gap between my departure and Thelma's arrival, thereby avoiding a condition known in professional circles as cage fatigue. If it happened that I was running late and forgot to bring the set into the dining room, Boris would scream his indignation at such a pitch that I never got farther than the elevator with my memory lapse.

Boris's taste in morning media ran to action shows. Shoot-'em-up westerns were favorites. When the posse rode for the bad guys with guns blazing, Boris would slap his thighs and jump up and down. He enjoyed commercials, especially those showing food or beverages, and "uh-uh-uh"d when anything looked familiarly edible. (When he was out of the cage, he'd rush to the set and reach for the tall, frosty glass of whatever was being poured, convinced it was Yoo-Hoo.) Talk shows bored him, cartoons baffled him, but Jack LaLanne's exercise program fascinated him. We were never sure what it was about the program—LaLanne's voice, the movements, the fact that it was broadcast on our clearest station—but Boris would watch that program more attentively than any of the others; so attentively that after three months of viewing he was actually doing some of the exercises! He'd lie on his back and bicycle pedal, roll his hips from side to side, then bring his legs back over his head, always in the same recognizable sequence. I'd thought of writing to LaLanne to tell him about one of his more successful pupils, but decided it might discourage an inestimable tonnage of housewives to know that an infant ape could out-slim them. Besides, Boris's individual leg lifts weren't so great.

* *Babysitter Blues* *

There were times, albeit few of them, during those early days when Jerry and I would have to go out for an evening. This posed major problems. Shep was too young to stay by himself, and if Thelma wasn't available we needed a babysitter. Not just any old babysitter, but one psychically strong enough to withstand Shep's electronic verbosity, one physically courageous enough to endure Ahab's volatile temperament, and one emotionally sensitive enough to love, feed, and diaper an ape. In a city of eight million people, there were exactly two: my mother (for whom coping was a way of life) and thirteen-year-old Jeffrey Gumprecht, for whom babysitting was a business. Being geographically desirable by virtue of living a few doors away, Jeffrey was our first choice.

Jeffrey was the son of a neighboring child psychoanalyst and, not so coincidentally, the owner of Ahab's mother. Because he'd known Ahab as a pup, he was the only human being outside our immediate family who could put his face close to Ahab's without sacrificing his nose. Jeff liked Shep and had the distinction of being one of the few people who could converse with him fluently on whatever beam he was tuned for. As for loving, feeding, and diapering an ape—well, there was always a first time.

The first time Jeff came to sit for our menagerie, we were as nervous as new parents. Jerry told him Ahab's walk times (if you thought Boris was upset when his routine was broken, you should have seen Ahab!) and showed him how to maintain the best grip on Ahab's leash. I explained how to prepare Boris's formula, wrote down feeding hours, diapered my pocket book to illustrate the proper procedure, and left a list of telephone numbers that covered an entire sheet of legal paper and every emergency. Jeff took it all in with pubescent equanimity; he was a cool kid. We blessed him, and meant it.

When we returned home later that evening, Jeff was dozing peacefully and unscarred on the couch.

"How did it go?" I asked tentatively.

"Swell."

"No problems?"

"Well, there was one, but Boris solved it. I diapered him, made his formula, but then I wasn't too sure how to go about giving him his bottle. I mean, I didn't know if he should be lying down or have his head raised or what. Anyway, while I was thinking about it, Boris grabbed the bottle, gripped it with his feet, and drank the whole thing by himself." Jeffrey paused, then gave me a serious, manful look, a look I felt his father must have given to parents of some of his patients. "Maybe he's ready for more autonomy," he said. "You could stunt his emotional growth by doing too much for him, you know."

I told him I'd be on the lookout for that pitfall and he seemed relieved. I spent a restless night reviewing my behavior with Boris, wondering whether I really had what it took to pass Spock-Freud-jungle muster.

On nights Jeff wasn't free, we rang up our second choice. Mother would always arrive with a cheery disposition and have something nice to say about all of us. (She did not always leave that way.) She'd smile into Ahab's barks, saying, "That's a good boy, good boy!" as if he was a docile basset hound who'd rolled over to be scratched, and she'd praise whatever Shep came forward to exhibit, be it a homemade transistor radio or a chain of paper clips, with rapturous enthusiasm. As a model and an erstwhile actress, mother moved through apartments, crowds, subways, and life as if she was being followed by a spotlight. Boris's exoticism delighted her. She kept his photo in her wallet, regaled friends with anecdotes about her new "grandson," and openly viewed him as a ticket to stardom. No matter how often we tried to explain that Boris wasn't trained to be a show chimp and that we had no intention of making him one, mother was convinced he could further her career. I couldn't visual-

My mother the actress and Boris the ham: the spotlight wasn't big enough for both of them.

ize agents clamoring for a sixty-year-old woman and an ape, but mother would always counter with "You never know about these things." The trouble was, her fantasy of being a dowager jungle queen was far afield from her real relations with Boris. Though she loved him dearly, it was obvious she preferred the abstract to the real. Try as she would to relax, whenever she'd hold Boo in her arms it looked as if she was attempting to pass through customs with something illegal. She'd talk rapidly, grin nervously, and ask us to take him from her when he reached for her hair. Boris was quite fond of her, adored her silver white hair, and was often reluctant to part company with his "grandma," which never helped put mother at ease. There is something disconcerting about an ape that won't let go of you. As a rule, we fed and diapered Boris before we went out, when mother was sitting.

Needless to say, when Jeff or mother wasn't available, neither were we.

* *Social Clamor* *

When you acquire an ape, unlike buying a TV, a new car, or tropical fish, you become aware of a curious upsurge in popularity. Once word of Boris had gotten around, we were deluged with calls from friends and relatives wanting to visit. Neither Jerry, Shep, nor I was antisocial (though I can't say the same for Ahab), but we fretted about exposing Boris to an onslaught of germs. Chimps, as we'd learned, could contract almost any human disease, and I had visions of Boris coming down with all of them. I began to put off callers with outrageous lies: "We're repainting and the place is a mess," "Shep has funny red spots on his face and we're not sure what it is," or "Jerry's crashing on a book and won't see anyone."

"We can't stall them all forever," Jerry said, not illogically.

"I know."

"Look, newborn humans catch diseases, don't they?"

"Of course."

"Well, people came to visit Shep when he was born."

"That was different. He slept most of the time. No one disturbs a sleeping baby. Boris jumps around and looks adorable. Everyone's going to want to hold him, play with him, sneeze on him."

"I'll tell them to have a complete physical before they come over," Jerry suggested.

"Don't be absurd."

"Look who's talking."

I gave in, and the hordes descended.

People's reactions to Boris and his to them were an education, or at least that's one way of putting it. Boris seemed to delight as many visitors as he unnerved. Very few could simply relax and enjoy his antics. They felt compelled to respond to him but weren't sure how. They'd joke with him as if he were human and then look embarrassed about it because they knew he wasn't, or they'd sort of pet him like a dog and then look embarrassed about it because he acted too human. Boris couldn't care less. He adored the attention and would go to great lengths—our floor-to-ceiling draperies, our pole lamp—to ensure he had it all. I soon realized that he and my mother would never make it as a team; the spotlight wasn't big enough for both of them.

Boris was quite forward with guests he liked.

Another aspect of ape behavior which many people seem to find enjoyable is the ingenuity that they show when attempting to escape, and the havoc they cause while at liberty.

Men and Apes

Nothing pleased Boris more than company. The little devil was a crowd pleaser, and he somehow knew he could get away with mischief when there were other people around. Besides, as he rightly assumed, the fans expected it. Our guests would actually encourage Boris's wildness. One couple thought his dropping-books-from-the-shelves routine was the cutest thing they ever saw, which didn't say too much for their range of experience (or our roster of acquaintances, for that matter), and when I pulled Boris down and scolded him, they made me feel like a tower guard at Auschwitz.

For some guests, befriending Boris was a point of honor, something like a primitive puberty rite. This always surprised me. My feeling was that if you got past Ahab, you'd already filled your macho quotient for the day. But then there were people like Seymour Geller.

Seymour Geller was a neighbor we didn't really know, didn't really like, but who'd found my wallet in the elevator and considered its return a guarantee of lifelong friendship, or at least worth an invitation to bring his daughter, Dawn, to see our chimp.

What could I do? There had been thirty-five dollars in the wallet, and none were missing. I told him Dawn could meet Boris on Sunday.

On Sunday afternoon, Seymour and his daughter arrived. Dawn was about six or seven, naturally shy. She looked as enthusiastic about meeting an ape as she would about playing jump rope with a python. She was still in shock from Ahab's barks when Seymour towed her to Boris's cage.

"There he is. What did I tell you?"

Boo jumped to the chain links and launched a series of vociferous "ho-hoo-uhoo"s, his assertive greeting to strangers. The poor child went rigid with fear.

"There's nothing to be afraid of," Seymour said reprovingly.

Dawn didn't agree. She tried to pull back, but Seymour held her there. Boris thrust his arm through the wire.

"You see? He's trying to make friends. Shake his hand."

Dawn shook her head.

"Go on, honey. He won't hurt you. Look." Seymour pumped Boris's hand masonically. "See? He's nice."

Dawn was not convinced.

"A lot of children are apprehensive about animals," I said, knowing that when you attempt to advise parents about their progeny you're treading a precarious line. "Maybe it would be better if—"

"Maybe it would be better if he was out of the cage," Seymour said. "I think the cage is frightening her. She associates cages with wild animals."

"Chimpanzees are wild animals."

"You know what I mean."

I knew what he meant and realized the impossibility of convincing him of what I meant. I reapproached the situation. "He's really very tired. He was running around the apartment all morning. Why don't you just give him a marshmallow. How's that?"

Seymour thought it was a splendid idea; Dawn wriggled as if nature's call was imminent. I handed the marshmallow to Seymour.

Boris saw the transaction and started hooting. He stuck his hand through the cage and let it dangle, palm up and ready.

Seymour uncoiled his daughter's fist and pushed the treat into it. "Just give it to him."

Dawn stepped forward and extended her hand, staring at it as if for the last time. Boris plucked out the marshmallow, popped it in his mouth, and reached out for another.

The little girl was so stunned by what she'd done that she just stood there with her arm outstretched, flexing her tiny fingers incredulously.

"There. Wasn't that fun?" Seymour asked.

I didn't hear Dawn's answer.

Seymour thanked me for letting them visit and hoped they could come again another time. I said I hoped so too—and crossed my fingers behind my back. I silently resolved to be much more careful about my wallet in the future.

Boris had an eminently larger social tolerance than either Jerry or I, though he did have preferences. He was a ladies' man. A scientific study done by the Washington Primate Center showed that male chimpanzees started very young to investigate females of their species, but the study noted that they never made advances toward their own mothers. An unscientific study done in our apartment showed that males started very young to investigate females of other species and didn't exclude their mothers. Once Boris walked into the bedroom while I was undressing, and it upset him greatly. He stared, touched me tentatively, and "hoo-hoo"d with consternation, quite concerned about what he appeared to envision as my inexplicable depilation. As soon as I'd dressed, he relaxed, but I often wondered whether he had certain reservations about me after that.

When a group of men and women happened to be in the apartment, Boris would always head for the distaff side first. He'd sniff their shoes, mouth their ankles, trail his fingers unashamedly up their stockinged legs, and behave altogether like an ardent foot fetishist in an X-rated movie. This was, of course, due in good part to the fact that he was only two feet tall.

> At an early age the subhuman primate often evokes all the maternal responses normally given to a human infant.
>
> *Men and Apes*

Women were constantly fawning over Boris. They cuddled him, scratched his back, gave him bracelets and necklaces to jingle, and cootch-coo'd him into soft, pant-hooting ecstasies. On the other hand, men were always wanting to throw him up in the air and wrestle him. His sexual preference did not seem unreasonable.

I began to relax more about having company and only on occasion regretted an invitation I'd made. The occasion was always one in which the guests, whether by nervousness or nature, made deprecatory comments about Boris. The upshot would be that Jerry and I would make hostile deprecatory comments about them. I vowed that the next person to come into the apartment and tell an off-color ape joke was going to get punched in the nose. Jerry saved a lot of noses by diplomatically telling people, "Yes, we've heard that one before," and changing the subject.

* *The Suspended Banana* *

For all Boris's enjoyment of company, he was really happiest when surrounded by his regular tribe. He particularly enjoyed playing with Shep and definitely treated him differently from an adult. One of Boris's favorite activities was to stand erect on Shep's lap and "groom" him, carefully looking through his hair. Jane van Lawick-Goodall has pointed out that as infant chimps begin to spend more time away from their mothers, their need for reassurance is met in physical play with other youngsters and in much social grooming. As they grow older and their playtime decreases, their grooming time increases. Mature adults may have a grooming session that lasts as long as two hours. (Shep would never sit for it, of that I was certain.) Also, what chimps pick off each other are usually small bits of skin called *scurf,* which they eat. Supposedly, the scurf, saturated by body oils and irradiated by the sun, provides them with the vitamin D they're deprived of by their hair, which limits the penetration of ultraviolet sun rays. I never saw Boris eat anything that he found in Shep's hair, but that could have been either because he was getting vitamin drops every morning or was just a sensible chimpanzee. Dr. Geoffrey H. Bourne, Director of the Yerkes Primate Center, has said that if an animal grooms another animal,

Boris and Shep goofed for hours with grocery cartons.

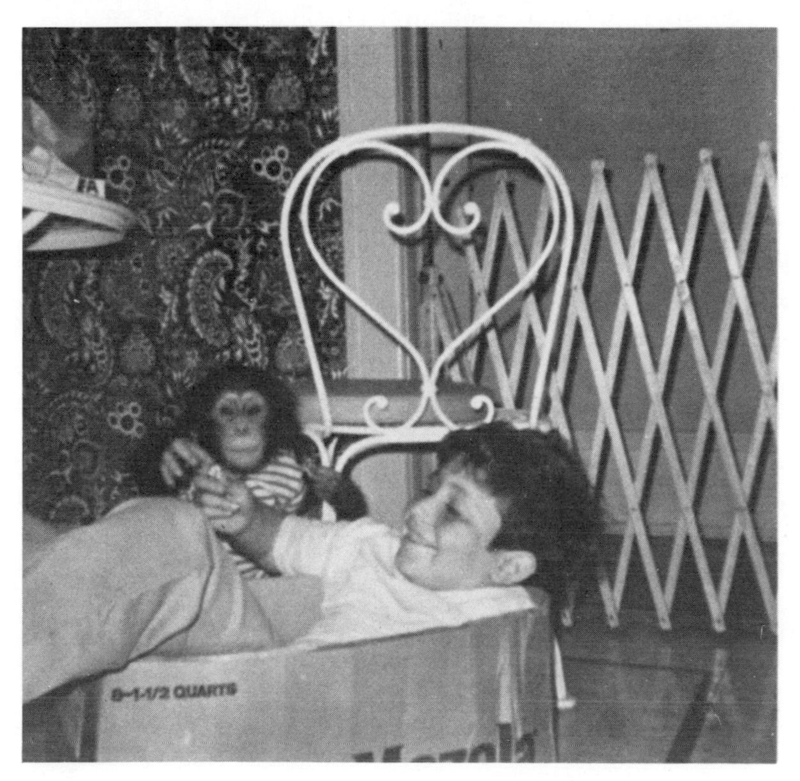

it means the "groomee" has been accepted by the groomer, so as far as Boris was concerned, Shep was aces!

The two of them were quite a pair. Shep managed to bridge any anthropological gap with games that appealed to him as much as to Boris. Grocery cartons from the supermarket were more fun for those two than anything a toy manufacturer could devise. Usually, Boo would sit in his cardboard limousine, smacking the sides, while Shep towed him recklessly around the apartment. Ahab became the movable obstacle as they went coursing from room to room. He jumped, barked, pulled Boris's blanket from the cage, and shredded it into pieces of cloth no larger than scurf.

Shep would also try to play educational games with Boris. Having seen a photo of Wolfgang Köhler's chimps building a pyramid of boxes to reach a suspended banana, Shep wanted to re-create the experiment. Boris was too young, or too lazy, to do anything with the cartons except butt them around the kitchen with his head, so Shep planned to construct the pyramid himself. The hard part was suspending the banana. Our ceiling was nine feet high, and Shep was just a bit more than four feet tall. He theorized a rather clever system of pulleys and drew me a detailed schematic diagram that ostensibly proved it could work. The diagram looked frighteningly incriminating, like a plan pilfered from an aerospace program. I didn't understand it at all, but when Shep explained that it entailed drilling a hole through the kitchen ceiling I quashed it. I was willing to sacrifice a lot for science, my child, and my ape, but that did not include my kitchen ceiling. We compromised by tying a string around the banana, tying one end of the string to a stick, and letting the stick jut out of one of the cabinets like a fishing pole. Boris thought the whole thing was terribly amusing. While Shep carefully piled the cartons on top of each other, Boo climbed a chair, stepped on the counter, and, calmly walking to where the banana was dangling, swatted it to the floor. So much for Wolfgang Köhler.

CHAPTER 6

* *Swinging Style* *

Inspirations in our household were often profuse, rarely profound, and always to be followed through with utmost caution. Impetuosity was for us like an old family curse; it never failed to bode ill.

One Saturday, after a particularly uneventful week, Jerry stood up from the breakfast table and announced that he was going to renovate Boris's cage. Before I could even suggest leaving well enough alone, he had his tape measure in hand.

"Shelves!" Jerry said, and was off and humming. Boris, he explained, needed shelves of his own—to lounge on, to jump on, to express himself arboreally on. Jerry was going out that very moment to buy the lumber. I shuddered inwardly with a prescience born of our sole family forte: hindsight.

When Jerry returned, he was carrying two excessively expensive boards and, for a reason that still escapes me, a can of bright green paint. (Our dining room was painted a lovely subdued gray.) I believe Jerry felt that Boris would be happiest with a "jungle" color, and who was I to say he wouldn't?

Once again, our dining room rug became plush sawdust. Shep and I curbed Boris's appetite for delectable nails and screws by diverting him with games and marshmallows while Jerry ran through his litany of standard carpentry obscenities. On Sunday, the first slightly sticky shelf was mounted, and Jerry couldn't wait to whelm Boris. Anticipating ecstatic somersaults and hoots, he rushed Boo back to the cage. Unfortunately, Boris's reaction was less than par. Boo took one look at the flat green protuberance, touched it with one finger, and decided that it was definitely something to keep away from. He crawled into the corner of his cage and stayed there.

Undaunted, and determined to make Boris happy whether he liked it or not, Jerry mounted a second shelf. This one was larger and higher up than the first, though just as green. Boris hated it. He didn't even climb close to it, but simply trembled and huddled down in the safety of his bed.

"He probably just has to get used to them," I said gently that day . . . and the next . . . and the next.

Jerry was no more consoled than Boris, who absolutely refused to leave the floor. Finally, on Friday, as Jerry was staring morosely at Boris, who eyed him in the same manner, the solution surfaced.

"I've got it," Jerry said. His teeth were clenched and there was such savagery in his tone that it was not altogether unreasonable that I feared he was contemplating some act of violence.

"We need a pole."

"What for?" I asked suspiciously.

"A pole to join the two shelves," he said. "It will give Boris a handhold. He'll feel secure. He'll love it."

I crossed my fingers as Jerry dashed out to the lumberyard. When he returned, he dragged out the can of jungle-colored paint and immediately applied it to the pole. He propped it against the wall to dry. It looked like Jack's beanstalk.

"Once it's in, he'll love it," Jerry assured me.

I was convinced, but I had my doubts about Boris.

Late the next day, after much hammering, cursing, and testing, the pole was firmly mounted in Boris's cage. Jerry stepped back as I brought Boris forward. It was the moment of truth, and I wasn't sure I was ready for it. As Boo entered the cage I held my breath, waiting like a Colombian native for El Exigente's verdict.

Suddenly Boo grabbed the pole and swung himself up on the first shelf, hooting loudly, happily. Jerry kissed me. El Exigente approved. The natives were happy.

Boris quickly established a close working relationship with the pole. He swung round on it, climbed from shelf to shelf with it, and clung to it when he sat watching television.

"Now," Jerry said, "what he needs is a swing."

"Why don't we quit while we're ahead?" I suggested.

Jerry gave me a look that made me feel an inch above a child beater. I told him I'd go to FAO Schwarz on Monday. FAO Schwarz is *the* toy store in New York. It stands on Fifth Avenue in the shadow of the Plaza Hotel and caters to a clientele that will pay one hundred dollars for a stuffed animal that would cost less than half the price live. I could have gone to any one of the numerous discount toy outlets around the city for Boris's swing, but I was feeling guilty. Guilt always makes me spend more than I ought to.

I'd made up my mind before I entered the store that I was going to get Boris the best swing there was. (I'd had visions of an endless selection of swings in all widths, strengths, and colors.) It wasn't difficult. There were only two swings in the store. One was attached to a combination slide-seesaw-rocking-horse-monkey-bar unit that cost six hundred dollars. The other had two chains with hooks and a light plastic seat. It was nine ninety-five. I recognized it instantly as the best swing there was.

The salesman informed me that it was nontoxic, hazardproof (the soft plastic seat had been tested by Consumer's Union and rated above average in minimum damage done to children's heads), rust resistant, easy to install, and durable. All for nine ninety-five? I was thrilled. I thought about buying two, three. We could festoon the apartment with them. I was certain I'd lucked onto the bargain of the year. I controlled myself; grew skeptical.

"Is it really durable?" I asked.

The salesman rapped his knuckles against the seat. "Your child will outgrow it before it wears out."

"It's for a chimpanzee," I said.

"I see," he said, with no discernible ruffle to his composure other than two swift eye blinks. He rapped the seat again as if to double-check himself. "I'm sure it will be fine."

"I'll take it."

"Can I help you with anything else for the—um—chimpanzee?"

I hadn't really thought about it. "Like what?"

"Toys or games. Not Monopoly, perhaps, but something like Chinese checkers might be nice."

"He's not even a year old."

"Hmm. An abacus perhaps? Lovely colored beads. I've heard mothers say that it kept their child busy for hours."

"His attention span is short."

"Oh. Well, just a thought." He looked disappointed.

"How much are they?"

"We've a marvelous one—very sturdy—for twenty-three fifty."

Twenty-three fifty! For beads! I could buy two and a half swings for that. I thanked him very much for his help and told him that I'd stick with what I had, that I didn't want to spoil the chimp with too many presents. He said he understood, telling me in a small burst of confidence that he'd witnessed displays of parental indulgence that would make me sick. I said I could believe it, and went to the wrapping desk feeling righteous.

I couldn't wait to get home and witness Boris's delight when he saw the swing. Jerry was just as excited as I was and stopped his work to install the swing. Boris watched the proceedings suspiciously, and when we let him back in the cage he ignored my

FAO bargain as if it wasn't there. We tried pushing it, jingling the chains, but no use. He climbed on his shelf and stared at the wall.

"I don't understand it," I said, feeling betrayed by every Tarzan movie I'd ever seen, and by Consumer's Union. "It's unnatural."

"Maybe it's the seat?" Jerry said. "It has those odd holes in it."

"It's special hazardproof, nontoxic plastic," I said loudly.

Boris was not impressed.

"Let me try something," Jerry said. He removed the swing and replaced it with two ropes and a single bar. It was an elegant trapeze, I had to admit. "Now let's see what happens," he said proudly.

What happened was that fifteen minutes later I was screaming, Boris was squealing, and Jerry was frantically unwinding rope from around Boris's neck. The elegant trapeze was not hazardproof. It had to go. We reinstalled the swing, assuring ourselves that Boris would use it when he was ready, and gave him the trapeze bar to play with. We had no idea at that time that the trapeze bar would soon become his favorite possession. Within days he was using it as a tool and a weapon.

As a tool, the bar had a somewhat dubious function, pushing raisins around on the shelf. As a weapon, it was a veritable cudgel. Boris whacked the bars with it when he wanted attention, smashed the swing with it when he wanted noise, and clomped the shelf with it when he wanted out.

> Wild chimpanzees, Dr. Kortlandt said, definitely used weapons. They brandished or threw branches and clubs. The natives, when they said that the apes could catch spears and throw them back, were not really exaggerating.
>
> *On the Side of the Apes*

The bar, we decided wisely, was off-limits outside the cage.

* *The Great Water Discovery* *

Discovery is a wonderful process. It imparts a special delight unequaled in life by anything—except perhaps a piece of really good cantaloupe. Boris's discoveries of himself and his urban world were as exciting for him as they were for us, though sometimes he definitely had the edge. One of those times was when he discovered water.

> The general response of chimpanzees to water is universally agreed to be one of avoidance and even fear.
>
> *The Apes*

It began with Ahab's water dish. One day when Boris was ambling around the kitchen, he put his finger into the bowl, withdrew it, and watched with fascination as the drop dripped to the floor. Although Boo was used to being washed in the bathroom, he was usually too busy grasping for other things to pay much attention to the water. But this was something else. Balboa had seen oceans before the Pacific, hadn't he? Ponce de Leon had seen fountains before he hit Florida, hadn't he? Ahab's water dish was no less a discovery for Boris. Suddenly a whole new aqua world opened up.

Boris would dip his hand into Ahab's dish and dribble the water into his mouth. In the wild, this behavior is sometimes called finger-drop drinking. In an apartment, this behavior is called very messy. Before we had a chance to share in Boris's joy of discovery, we were at our wits' end trying to keep dry. Dipping led inevitably to tipping, which meant a daily washing of my kitchen floor whether it needed it or not. Tipping led to splashing, which often meant a mid-evening change of clothes for Jerry, Shep, or me. And splashing soon led Boris into the bathroom, where the toilet was all too accessible. Now, no longer were his wash-ups merely difficult—they were splashy, soggy, and relentlessly sloppy. If he wasn't covering the faucet spout with his finger, spraying everything in range, he was turning on handles as soon as I shut them off. We yelled, we shouted, we "No! No! No!"d him until a barrage of broomsticks pounded from apartments above and below, but to no avail. Shep asked if he could sleep with an aqualung, in case Boris escaped in the night and drowned us in our beds. Thelma asked if we could buy a new mop. Jerry asked if I'd read about anything like this in any of the books I'd bought. My answer to all of them was still another shrill no!—though I did give in on the mop. I was in tears. I feared pneumonia for Boris, mildew for the apartment, and arthritis for the whole family. I hadn't any solution to the problem, other than confining Boris to the cage, which I knew I could not do. Then the miracle happened.

Miracles are funny sometimes. They can creep up on you so slowly that you don't realize they're miracles at all. That's the way this one occurred. As sneakily as Boris's water crescendo had mounted, it receded. Day by day he became less and less interested in converting our apartment into a pool. I realized it as our paper towel supply

No one had ever told Boris that chimps don't like water.

When Boris cut loose in the kitchen, we had a ladder in readiness to curtail his high-minded ideas.

began lasting longer and longer, as our bathroom carpet ceased to squish underfoot. Although he did pull an occasional water prank once in a while, his aqua obsession disappeared. I chalked it up to "a phase," and gave thanks.

* *Apron Strings Untied* *

Every time Boris outgrew or tired of some maddening activity, I gave thanks, but almost always my relief was premature. No sooner had he mended one set of mischievous ways than he created others. With Boo, I never knew when I was well off until it was too late to reverse the outcome. The most intense example of this phenomenon was when he finally severed the apron strings that had previously convinced me I was doomed to walk through life with a chimp on my shoulder, or hip, or leg, or back. It happened, ironically, on Mother's Day.

I was so used to the invisible umbilical cord that joined us, I'd become confident that no matter where I went in the apartment Boris was sure to follow. Jerry and Shep had gone to buy a cake, and I was in the bedroom putting away the present they'd given me, when I realized that Boris was not in sight. I called, but still he didn't come. It was more than odd; it spelled trouble, and I knew it.

I walked slowly back through the apartment, calling his name, telling myself there was nothing to worry about but secretly dreading the worst. Fortunately, the windows were closed, Ahab was sleeping peacefully, and the bathroom was empty, allaying my three prime apprehensions. Then I heard a loud "hoo-hoo" from the kitchen. Boris had climbed the steam pipe to the ceiling and was happily bombing the floor with bits of banana. But the steam pipe was adjacent to a large kitchen window that I had cleverly, I'd thought, opened safely from the top. It now offered tempting access to unknown delights. I closed it immediately and held out my arms.

Boris "hoo-hoo"d energetically and lowered one of his hands toward me, but when I grabbed for it, he pulled it out of reach. I tried again, with the same result. He had discovered a new game and was practicing winning.

All right, I thought, let him stay there—until I noticed my teapot on top of the kitchen cabinet. The teapot was delicate Bavarian china, my sole family heirloom. I'd guarded it like a secret love for years. I had kept it on top of the cabinet because that had seemed the safest place in the house. But now it was dangerously within Boris's reach. I knew if I climbed up to get it, Boris would think I was coming for him and bolt. And the route offering the fastest avenue of escape was across the top of the kitchen cabinets, across my teapot.

I had to think fast. I tried to recall what I'd read about policemen who were trained to talk down jumpers from buildings, but my mind wouldn't budge. At a loss for anything inspired, I reached into the refrigerator and pulled out a plum. Boris began a series of energetic food barks. I was getting warmer. Boris reached out his arm.

"Come on, Boo. Nice juicy plum. Come down and get it."

Boris simply "uh-uh"d louder, and every time I tried to grab his hand he whipped it away.

As my cajoling grew more elaborate, his obstinacy solidified. My fear for the teapot was rapidly being replaced by my anger at Boris. I tried the straightforward hard line ("Get down here this instant!"), the mock-excitement encouragement ploy ("Come on, boy. Come on down. That's it, that's it!"), and another serpentine food inducement ("Look, Boo, a beautiful banana"), but none of them

Why my hair turned gray.

worked. I decided to play my last card, the old-fashioned get-'em-with-guilt routine. I buried my face in my hands and pretended to cry. Through my fingers I saw Boris peering at me. His head was cocked to the side and his lips were pursed; he was concerned. It was an encouraging sign. I played the injured mother to the hilt. Within seconds, he was down the pole and tugging at my skirt, looking very distressed. I picked him up, continuing to feign unhappiness, which wasn't easy, considering my relief, while he "hoo-hoo"d anxiously. He touched his fingers to my eyes, my cheeks, and looked thoroughly contrite. Then he puckered his lips and kissed me. I melted faster than sherbet in sunshine. I snuggled him and let him down. He "crutch raced," knuckles to floor, out of the room and down the hall to goof with Ahab, and didn't even look back. I knew then that he was on his own and we were going to have to be ready for it.

When I told Jerry what happened, his face darkened. I was confused. The teapot was safe, Boris was independent, what was wrong?

"You realize what we'll have to do, of course," he said.

I realized nothing of the kind, and said so.

"We're going to have to buy air conditioners, unless, of course, you want the only sunlight we get to be filtered through iron grills."

I missed the connection. "What for?"

"We're not going to be able to open any window more than a crack, with Boris around."

"Oh."

Jerry asked me if that was all I could say, considering the price of air conditioners. I told him that the price of liberty, even for an ape, was never cheap. He didn't even smile. I saw the way the rest of the day was headed and curled up on the couch with a book.

CHAPTER 7

* *A Sunday in the Park* *

With summer almost upon us and the weather abundantly balmy, we decided it was finally safe to take Boris outside. It had been a long indoor winter, and it was time for him to get some fresh air and a change of scenery.

We phoned several friends and asked if they'd like a visit from Boris, thinking they'd be thrilled. We received polite refusals all around. We settled on taking Boo to Central Park.

The park was only two blocks from our apartment, and going there with Boris seemed a simple, quiet, nice Sunday thing to do. Unfortunately, we couldn't have been more wrong. The moment Jerry carried Boris out onto the street there were crowds. Kids, parents, delivery boys, and dogs—dozens of them—trailed us all the way to the park. Everyone we passed pointed or tried to pet Boris. They assailed us with questions, the most frequent being "Is that a monkey?" "Where's his tail?" and "Does he bite?" At first our answers were automatic, to the extent that when a hot dog vendor hawked out, "Hot dog or soda?" Jerry shook his head and shouted back, "No. Chimpanzee." After a while, we simply couldn't bear the relentless barrage.

We grew surly. We snapped hostile answers. When someone asked if he was a monkey we told them no, he was Charlton Heston. When they asked where his tail was, we said it was hanging from our automobile antenna. When they asked if he bit, we growled and thrust him forward. By the time we reached the park, even I didn't like what we'd become.

We found the most deserted spot we could, though it didn't stay that way for long, and sat down. We were exhausted, harassed, but at least our Boo was going to have some outdoor fun—we thought. Ha!

First, Boris wouldn't walk on the grass. Not only didn't it rekindle wonderful jungle memories, it made him scream out in fear. We tried to convince him that he really loved it, that he just had to give it a try, but he stood firm—on our laps! Well, we thought, he likes to swing, so why not take him to the monkey bars? The reason became evident only when we got there: Boris wouldn't touch them. We

Climbing lamps was one thing, but nothing could induce Boris up a tree.

Photo: David Sagarin

tried the swings, the slide, trees, but nothing worked. After two fruitless and frustrating hours, Boris was still clinging to Jerry's neck. Our first major outing was a bust; Boris was a drag.

On our way back, with Boo squinting against the bright light and breezes he wasn't used to, adding guilt to our already dejected state, we met a group of young hippies. They took one look at Boris and cumulatively fell in love. One girl handed him a dandelion, which he promptly and happily popped into his mouth, and she fairly swooned with delight. The group reaction was nothing less than awed reverence. As we were leaving, one of the young men asked where Boris slept. Jerry told him that he had a bed in his cage.

"I wouldn't keep him in a cage, man," the youth said. "He's real."

Jerry smiled. "Unfortunately, so is our apartment."

They shook their heads sadly as we left. We felt like doing the same thing.

When we got back to the apartment, we were totally exhausted. The fresh air we'd taken for Boris had knocked us out. But not Boris—no sooner did we put him in the cage than he was swinging and stomping around on his shelf, acting for all the world like a chimp who'd suddenly been set free in Central Park.

* *A Birthday by Mouth* *

For some reason, the appearance of a new tooth in a child's mouth is an important event, at least according to baby books, which reserve spaces for dutiful mothers to fill in the dates of each arrival. I recall that when I was a pram pusher there were many mothers who'd boast about their child's early production of a tooth as if he'd laid a golden egg. I personally had never paid much attention to the time, order, or rapidity of Shep's dentition. I had confidence that he would not be a toothless toddler and failed to see any reason for concern or more than passing interest. But Boris was another matter entirely. First, I had reason to be concerned. Any tooth of Boris's that went uncounted was potentially a concealed weapon. And second, I felt that the order and appearance of his teeth would more accurately pin his age. The trouble was, the appearance of his canine teeth contradicted the age the Animal Medical Center had given us. He was either four months older than we thought or dentally precocious. Jerry and I opted for the latter, because to accept the former would mean canceling the birthday party we'd already compromised on by giving and taking a month in either direction.

But it bothered me. Four months in the age of a chimp or child counts a lot. How was I to evaluate him or plan his diet? I decided to call the Bronx Zoo and fix Boris's birthday once and for all.

The Bronx Zoo, or more formally, the New York Zoological Park, covers two hundred fifty acres and an extravagance of species. Their primates are as varied as they are abundant, and I felt confident that some member of the Zoological Society staff would be able to give me the handy rule of ape age detecting. Surely there was something simple, like counting the rings on redwoods. So I rang up.

The switchboard at the Bronx Zoo protects its keepers and staff with leonine ferocity. Getting a call through to the primate section was about as easy as reaching the President. After I'd convinced the woman that I was not a nut, not inquiring about park hours, circus rental, or a missing child, she reluctantly put me through, or tried to.

"They don't answer," she said, almost smugly.

"Well, would you keep trying, please," I said.

There were two more rings and then the operator cut in again. "They still don't answer."

"When do you suggest I call back?"

"I really wouldn't know."

"Well, how can I reach someone in the primate department?"

"I really wouldn't know. I can try another office."

"I'm calling about a chimpanzee."

"Oh. Then you want primates."

"That's what I—" Before I could finish, there was a ring and a woman picked up the phone. She gave a name that sounded like Tyrannosaurus; I dared not take a stab at it. I said a friendly "Hi!" and started to explain the problem. I'd gotten no further than "my chimpanzee" when she interrupted.

"Macaque," she said brusquely.

"Excuse me?"

"You have a macaque. They're short-tailed monkeys. Most people mistake them for chimpanzees."

"Listen, there's no mistake. He's a chimpanzee, believe me. What I want to know is—"

"Everyone thinks they have a chimpanzee," she said tiredly. "Pet stores pass macaques off for chimps all the time."

"Look, I have a chimp!" I was indignant, angry.

"You're sure?" There was that teasing lead in her voice, the kind that prepared you for a sandbagging.

"Of course I'm sure." I was seething now. Here I'd phoned to ask a few simple questions, and I was being interrogated, Boris was being maligned. I felt like cabbing it up there and punching her in the nose.

"May I ask how much you paid for him?" she asked sweetly.

"One thousand dollars."

She conceded that I probably did have a chimpanzee, and reluctantly offered a few feeding suggestions. She was no help at all on pinning Boris's age. I thanked her with a pointed lack of sincerity.

I tried her diet suggestions—strained baby meats, egg yolks, and cereal—and Boris expressed his discontent for a week, keeping his mouth closed, slamming the jars to the floor, and hooting angrily at every mealtime. I returned to his regular diet, which was supplemented with vitamin drops, and decided to guide myself from then on with Dr. Spock. Boris thrived. As far as his age went, he was going to be a year old on June first whether he liked it or not.

The birthday party was a small family affair. My sister and brother-in-law brought champagne for us, biscuits for Ahab (so he wouldn't get jealous or hungry), and a miniature football for Boris. Shep supplied a jungle version of "Happy Birthday" that he'd synthetically created by taping a variety of electronic sounds. It sounded a lot like a flock of choking chickens, but Boris "hoo-hoo-hoo"d ecstatically when he heard it, and it took some tricky talking to convince Shep that two renderings of the tune were enough. We gave Boo a basketball, which was more than half his size and which

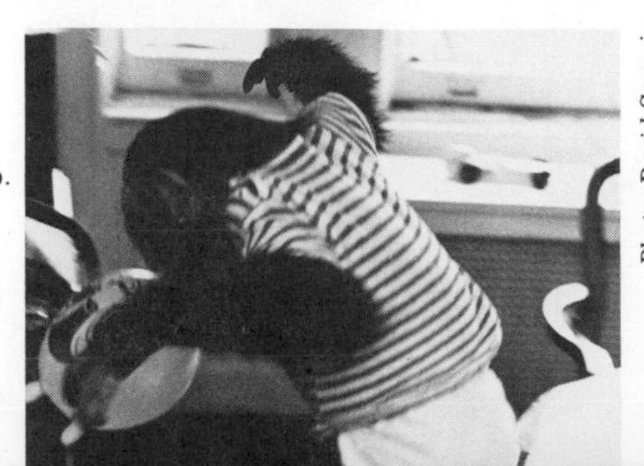

Boris would never walk, sit, or stand when he could leap. Unfortunately, there were very few times when he couldn't leap.

Photo: David Sagarin

he immediately rolled at a high speed into the sleeping Ahab's groin. At that point the partying escalated. Ahab chased Boris around the dining room, forcing Boo to leap to the top of the table for safety. Unfortunately, the birthday cake was right in Boris's line of survival. He was in it hand and foot before we could stop him. I consoled myself with the fact that he seemed to enjoy licking the icing off his toes. He also enjoyed champagne. My brother-in-law Ron offered him a sip and after the first taste he was hooked. That little mini-wino went cuddling up to everyone who had a glass and begged a drink. I hoped it wouldn't get to be a habit. Buying fifteen pounds of bananas a week was expensive enough.

The rest of the party was relievedly uneventful. Boris alternated draping himself over the basketball and sort of playing dead with pushing balloons into corners and kissing them. We thought it was a bit weird, but after all, it was his party.

When it was all over, Boo was supercharged with energy and barked angrily when we led him back to his cage. He expressed his discontent by pulling back his swing and hurling it against the wire mesh. Then he faced the wall, crossed his arms, and refused to turn around. Boris had postparty blues.

* *Walking Tall* *

Now that Boris was a year old, I began to fret about his being an underachiever.

> We distinguished eleven different types of locomotion in these highly adaptable chimpanzees, which were equally at home on the ground or in the trees. They were: quadrupedal walk, quadrupedal run, rapid run, gallop, vertical climb, bipedal walk, bipedal walk on legs with use of arms, ground leap, vertical leap, swing, and brachiation.
>
> *The Apes*

According to what I'd read and what I'd observed of Boris, he was behind his jungle cousins in both brachiation and bipedalmanship. I was quite content to discourage brachiation, mainly because the things he'd have to swing to and from, like our chandelier and draperies, were expensive and fragile. But bipedal walking was natural even in the wild. If a common, everyday, run-of-the-mill chimp could do it, why couldn't Boris? It was time for Boris to stand on his own two feet and walk like a man.

Actually, he had taken a few two-legged steps from time to time, usually when he was holding something, but even Ahab could do a short two-legged stand. I decided nature needed some remedial aid. I took to holding Boris's hand and walking him around the apartment. Now a six-room apartment is certainly comfortable enough to live in, but it is not set up for strolling. Six steps and you're in the bathroom, twelve steps and you're in the dining room, twenty-five steps and you've covered it. And there's nothing to see. I felt like a privileged convict, like a father-to-be pacing a small maternity ward, like a border guard for a very tiny country. Boris and I loathed every minute of it. He wanted to do his own thing and so did I, but we were chained together by my righteous belief in his well-being, which was unshakable, immutable. I gave it up in a week.

So he'd run on his knuckles for the rest of his life, so what? I was consoling myself with this thought one evening when Boris, who'd been amusing himself with a colander and a garlic press under the kitchen table, stood and walked across the room. I refused to be impressed. It was a fluke. Just because a baby says "Da!" doesn't mean he's started to talk. I was not going to be lulled into false euphoria.

But in the next few days, he bipedaled it a lot. He got a big kick out of it, taking four, five, and six two-legged steps and then securing himself with a knuckle on the floor or, when Ahab was in pursuit, reverting to his high-speed "crutch" run. He was proud of himself. Now he could hold a banana in one hand, a marshallow in the other, and still go where he wanted to. New vistas yawned before him and he became more assertive, much more sure of himself—to be quite honest, obstinate and sometimes downright ornery. I couldn't really get angry. I chalked it up to the "no" stage that children go through and assured myself it would pass. I expressed my encouragement for his new accomplishment by buying him a pair of sneakers.

Buying clothing of any sort for a chimp is always an adventure in creativity and courage. There's no way you can bring him into a store without subjecting both of you to ridicule and harassment. And going alone isn't a snap either, especially when you walk into a shoe store with a crude pencil sketch of what looks like the footprint of the abominable snow-child, which is what I did.

There were two salesmen in the store I entered, who looked like father and son. I decided to approach the son, calculating that his youth and eagerness to prove he was worthy of inheriting the business would make him more receptive to my rather unique problem.

"Excuse me," I said. "I'd like to buy a pair of sneakers."

"What size?"

"Well, that's the problem. I'm not sure." I flashed my most engaging smile as I took out the paper with the outline of Boris's foot. "All I have is this. Could you measure it and tell me the size?"

He looked at the paper, then back at me. The corners of his mouth twitched, unsure about whether or not he was being put on. "What is it?" he asked.

"His foot."

Now he was really undone. I didn't smile or laugh, and you could see he didn't dare risk offending me by saying what he was thinking.

"Well?" I asked.

"Er . . . how old is he?"

"About a year."

His face twisted in a sort of uneasy grimace. "I . . . uh . . . his foot is rather long and narrow. Have you had trouble fitting him before?"

"This will be his first pair."

"I see," he said, licking his lips nervously. "Well, I'll try, but I can't guarantee a perfect fit without an actual measurement." I noted that he held back from asking why I hadn't brought the child in or suggesting that I do so. He took the paper to the counter and pressed it down on the measuring rod. I saw his father, who was waiting on a customer, look up and transmit a stern, questioning look. The son ignored it and hurriedly told me that a size 6 might do it.

"Fine," I said, and watched as he tapped his father on the shoulder as he passed, indicating that he should follow him to the rear of the store. They huddled together and whispered. I studied the assortment of polishes and shoelaces in the glass case and pretended not to notice. When the son reappeared with the sneakers, I asked if I could return them if they didn't fit.

"Sure," he said, nodding more times than necessary. I had the feeling he thought I was dangerous.

As I left, the eyes of both father and son were almost palpable upon my back. Boris would never be Beau Brummell; I hadn't the fortitude.

I decided that the best way to spring my surprise present on Boris was to let him open the box himself. He had fun opening things—closets, drawers, blouses. Only the day before he'd amused himself for an hour by opening a twenty-five-pound bag of dog kibble and rolling the pellets around the floor.

He approached the shoe box cautiously, as he did all new things, his lips pursed. He lifted the lid and poked his nose inside. Then he took the top off and tossed it aside. He took out one sneaker, brought it to his lips, tasted it, put it down, and repeated the process with the other one. Then he took out the pink crinkly paper and raced happily across the room to play with it. I was unable to convince him that I hadn't wrapped his present in seven-dollar Keds.

But I was not discouraged. The pink paper kept him busy while I held him on my lap and put on the sneakers. But once the shoes were on, he forgot about the paper and watched me with what can only be described as mounting suspicion. As I tied the laces, I had to keep pushing his nose out of the way. When I finished, he sat perfectly still, his eyes fixed on me with the look of a child who realizes he's been tricked into a tonsillectomy.

"Okay. Here we go," I said, taking his hand. He slid off the couch and landed with a clump, two clumps. "Come on. Let's walk." I took a step forward. Boris did not move. He stared at his feet, which looked as if they were nailed to the floor. I took another step and tugged gently. Boris chugged into motion. Clump. Clump. He looked like Frankenstein's monster and moved with all the grace of the Heap in chains. He wouldn't walk at all by himself, wouldn't even crawl. If I didn't tug him, he'd just drop down where he was and stare at his feet. The pictures on the side of the box showed a group of children racing up a hill. I showed the box to Boris, pointed to the white sneakers all those happy children were wearing, and pointed to his own sneakers. He simply stared dazedly at his feet and remained motionless. Boris was not ready to become one of the gang. I took off the sneakers and felt like the good fairy who had broken a spell. Boris tore back to the couch and picked up where he'd left off with the pink paper. The next day I gave the sneakers to the mother of a three-year-old boy across the street. I told her they were samples I'd gotten in the mail. She was very pleased. Soon her little boy was going to run and play in his white sneakers like the kids on the shoe box. I consoled myself with the thought that Boris would probably never have gotten into group sports anyway.

* *Games Primates Play* *

Boris liked playing his own games, basically because he liked winning. And Boris rarely involved himself in anything in which he couldn't triumph. Ahab proved to be Boris's greatest playmate. Whereas Shep, Jerry, and I would tire of romping, and Thelma had no time for it, Ahab could always be counted on.

Boris and Ahab definitely had a relationship, though the exact nature of it still remains unclear. We could never be sure, for instance, if Ahab simply liked Boris, felt responsible for him, or hated him. Often Boris would run to Ahab, throw his arms around his neck, and hug him, but whether this was meant as an embrace or an attack was also uncertain. Boo, wisely, would split before Ahab had a chance to interpret it. But in many ways, Ahab was like Boris's big brother, and like all younger siblings, Boo emulated him. Regrettably, one of the ways he did so was barking at strangers.

> A chimp is capable of making 32 distinct sounds, all cacophonous, ranging from disgusting grunts of approval through moans of worry, piercing screams of anger, or roars or growls when enraged. He also can hoot, bark, whine or howl.
>
> *A World Full of Animals*

After Boris had been with us for six months, we took to forewarning visitors of the fanfare they'd receive. To walk into our apartment cold was like plunging into a stereo megawoofer gone amuck.

When Boris and Ahab played together, ice-cube hockey was one of their favorite sports. Ahab loved to chew ice cubes. Whenever anyone went to the refrigerator, Ahab followed and held them there until they paid him an ice cube. He'd usually nose it around the kitchen floor a bit for fun and then eat it; he ate about twenty-five ice cubes a day. The hockey game would begin as soon as Ahab began nosing the cube. That was Boris's cue. He'd scoot out from under the kitchen table, snatch the cube, and streak into the hall. Ahab would rage after him. Boris would then fling the cube to the other end of the hall, instantly diverting Ahab. He was really crazy about those cubes. Boris got so secure that he could wait until Ahab was almost upon him before tossing the cube. Sometimes, when he was feeling really macho, Boris managed double plays. He'd cut through the dining room and make it back into the hall to reclaim the ice cube before Ahab could locate it. Then he'd run past Ahab, sporting the prize, sometimes licking it, and fling it in another direction. Often, by the time Ahab got the cube it was no larger than a kernel of corn, but somehow it didn't matter. It wasn't how much he lost, it was how he played the game.

Boris knew he was the baby in the family, and much like human youngsters of similar status, he knew he could get away with plen-

Photo: David Sagarin

Sometimes Boris's plans to ambush Ahab would backfire.

ty, especially with Ahab. He'd often goad Ahab into barking and then race for the protection of my arms, hooting gleefully as Ahab was sternly reprimanded. Boo was so certain we'd protect him from Ahab that he often took malicious liberties.

The worst was his gonadal ambush. It would start innocently. Ahab would be walking around the apartment, minding his own business, and Boris would quietly disappear. Then suddenly Boo would spring out from behind a chair, dive daringly between Ahab's legs, swat the dog's testicles, and quickly leap out of jaw shot in a single bound. Boris would then remain in the safety of whatever high ground he'd gained, while Ahab stormed the apartment. But no sooner had Ahab calmed down and forgotten the incident than Boris, if we didn't intercept him, would be as ready as ever to launch another attack.

Perhaps the most fun they had together was playing football. Boris would walk up to Ahab waving his miniature football under the dog's nose. Ahab would lunge for the ball, take it from Boris, and run around with it. Then he'd bring it back, drop it on the floor between them, and wait. The object was to see if Boris could get the ball in his hand before Ahab got Boris's hand or the ball in his mouth. They'd circle the ball, feint at it. If Boris managed to snatch it, he'd run for cover and hurl it away from himself (note the similarity to ice-cube hockey) when Ahab got too close. Ahab truly enjoyed the game, for he liked nothing better than to prove his mouth was quicker than anyone's hand. Most of our guests accepted this on faith, especially when Ahab would drop a chewing toy on their laps and dare them to beat him to it. Boris was a lot braver than most humans when it came to challenging Ahab. But then, not many of our friends could scale a pole lamp in four seconds flat.

Boris preferred playmates, but he had pretty good times by himself too. One of his most pleasurable solitary pursuits was Beat the Sock. This was an ingenious game that required skill, coordination, courage—and a sock. Boris would take a sock and climb almost to the top of his cage. He'd hang there for a long moment, look down, look up, make quick calculations, then hurl the sock upward and simultaneously release his hold on the cage and drop to the floor, trying to Beat the Sock. He'd play the game over and over again. Final-

At nine months, Boris could spiral to the top of our living room lamp faster than stripes could swirl on a barber pole.

ly he got to the point where he could not only Beat the Sock, but catch it! It was a show-stopper.

Boris also did impersonations. Although his Winston Churchill left much to be desired, his Quasimodo was superb. But his best by far was, ironically, Tarzan. He'd hang by one arm and bring his feet up under his shirt and pull it down until it was off one shoulder. We thought it was clever and adorable, but couldn't know that a year later this adorable little feat would be driving us mad.

Photo: David Sagarin

We wondered if Boris's playing Tarzan indicated an identity problem.

CHAPTER 8

* *Once Around the Park Again* *

Toward the end of June, the sounds of laughing children enjoying the outdoors had magnified our guilt beyond endurance. Fully aware of what was in store for us, we decided nonetheless to give Boris another outing in Central Park. We phoned a photographer friend, David Sagarin, and asked if he and his wife Judy would like to come along. We guaranteed them that it would be an experience, though refrained from specifying what kind. Being adventurous and recognizing a strain of anxious hope in our invitation, they agreed.

When David and Judy arrived, Boris was revved up and ready to go. He'd been charging around the living room, jumping on and off the couch, for an hour. We were sure that this time he was ripe for the great outdoors. David slung his camera over his shoulder and Jerry slung Boris over his. We were off.

Our second walk to the park with Boris differed from the first in only one respect: every harassing moment of it was joyously photographed by David. He wasn't bothered at all by the questions, the pulling, the tugging, the jeers. "Good shot," he'd say, with a sort of excited encouragement, forcing us to smile in spite of our rapidly ebbing patience.

For the first forty minutes in the park, Boris again resisted all efforts to get him to dig nature. Every time we put him down he shrieked, so loud, in fact, that more than one passerby shot us a look indicting us for animal abuse. I was getting paranoid. It was unnatural for Boris to behave like this and was very embarrassing. Whoever heard of a chimp that wouldn't climb a tree? that wouldn't play in the grass? We did, that's who!

I suspected the worst: we'd overcivilized Boris. Jerry suspected the truth: he needed time to get used to new surroundings. Just as I was about to suggest that we head back to the apartment, Boris hopped from my lap onto the grass. The crowd of ten that had hovered around us doubled magically. At first Boris just sat there, but soon the children who'd been quietly calling to him began to capture his interest. He finally stood up and walked to a toddler sucking on a pacifier. The two studied each other with equal

The urban ape. A portrait in Central Park. *Photo: David Sagarin*

fascination. Boris had never seen a human so close to his size. When the little boy reached out to pet Boris, Boris reached up to pet him. There was an outbreak of "ooh"s and ah"s. David snapped pictures like crazy. I was so pleased, I felt like one giant grin. Then all of a sudden Boris yanked the pacifier from the kid's mouth.

The little boy screamed. His mother, who'd turned away at that moment to call her husband, panicked. Boris gave a terrified hoot and tore back into my arms, the pacifier clutched in his fist. It took several awkward moments for the crowd to convince the mother that Boris hadn't harmed her son, and several more for me to convince Boris to relinquish the pacifier. All was resolved happily. When Boris gave the pacifier back to the little boy, several other youngsters stepped forward and offered to compensate Boo with an assortment of treasures ranging from a Tootsie Pop to a well-chewed, but, we were assured, still flavorful piece of gum. To accept something from one child and not another would be disastrous, and to accept all that was offered was ridiculous. I declined for Boris as politely as possible.

The crowds came and went all day, and soon Boris even had his own police guard. It happened while we were trying to cajole Boo into climbing a tree. Our reluctant ape was staunchly ignoring all inducements to move up in the world—jingly keys, marshmallows, raisins—when a motor-scooter policeman came by and informed us that it was against the law to bring a wild animal to the park. Fear-

Boris wondered what everyone was looking at in Central Park. *Photo: David Sagarin*

ing a hefty summons or worse, I blurted that Boris had been traumatized by a fall from a tree as an infant and we'd only brought him to the park in the hope of rehabilitating him, so he could live a happy arboreal life like other chimps. I have no idea whether the policeman believed me, because as soon as Boris walked over to examine the motor scooter it didn't matter. One blink of Boo's nut brown eyes and the letter of the law was forgotten; the cop was hooked. He stayed with us and warded off stray dogs and rowdy spectators for the rest of the afternoon. We never did get Boris to climb a tree that day, but David got some wonderful pictures and everyone, even Boris, had a fine time.

* *Ape Tripping* *

Now that Jerry and I were used to the crowds that Boris wrought, getting less apprehensive about the diseases he could catch, we began to take him more places.

One day, after many of listening to Shep plead, Jerry took Boris to school. The children in Shep's class had been instructed by the teacher that they were not to annoy Boris, that if they behaved themselves, Mr. Mundis might let them pet the chimp, *but only one pet per child.* Jerry learned this when he arrived. Shep whispered to him that he'd already promised extra pets to several friends whom

he'd identify by nods to Jerry when they came forward for their strokes. Unable to condone the favoritism, Jerry gave a short talk on apes to the class while Boris munched a large banana, and then spent the rest of the visit monitoring the petting session, allowing all the children extra pats. Boris loved it—as far as he was concerned, it was a grooming orgy. Shep loved it too, and his popularity soared. It was the only time he resented the imminence of summer vacation; he had the class presidency in his pocket.

Buoyed by the success of Boris's day at school and by our last trip to the park, we insanely decided to push our luck and take our entire tribe to the country for a weekend. My mother had a summer home on sixty-three acres of woodsy New Jersey land and had been pressing us for a visit. She didn't seem perturbed at all when we informed her we'd be arriving en masse, which proved only that she was no smarter than we were.

> It was considered unlucky in ancient times to see an ape on leaving home.
>
> *Men and Apes*

We planned to drive out on Friday night, not wanting to miss a minute of what we envisioned as a carefree holiday. I should have known better on Thursday night when I found myself still packing for this three-day excursion at two-thirty in the morning. Aside from our clothing, which not only covered the weather spectrum but provided for such eventualities as dining out and possible stains, there were Ahab's dog food, treats, and toys and Boris's shirts, overalls, sweaters, playthings, Pampers, bananas, marshmallows, Yoo-Hoo, and raisin bread. The sight of a couple packing a car on a Friday night is not unusual in New York City, but when you see them stuffing in a barking German shepherd, a balking eight-year-old, and an excitedly hooting ape, it's an eyebrow raiser!

The ride out, in heavy Friday-night traffic, was an endurance test, and all of us failed. In order of breakdown, Ahab was first. Unlike your typical family dog, the sort that hears the jingle of keys and races you to the car, Ahab had a gastrointestinal phobia toward automobiles. We attributed this to the fact that when he was a puppy, the only time he went in a car was to visit the vet. Negative conditioning, we reasoned, and attempted to reverse it by taking him on short taxi rides to the park. The trouble was, the rides were never short enough. Ahab could get car sick before we stopped at the second traffic light, and usually did. The ride to the farm was an hour and a half; a very long hour and a half; two rolls of paper towels long.

Shep ran a close second to Ahab, mainly because he was riding in the back seat with the dog and was responsible for clean-ups.

Jerry came third. Because I was driving, he held Boris and had to oversee the back-seat clean-ups. Before we reached Paramus, he was tighter than a tourniquet and cursing vociferously. No sooner would he turn his head to see how Ahab and Shep were doing than Boris would lunge for the ignition keys, the gearshift, or the cigarette butts piling up rapidly in the ashtray. I tried to laugh at it all and pretend it was a Disney movie but was not very successful, with tears of frustration streaming down my cheeks.

Boris, glutton for chaos that he was, got the most out of every maniacal minute of the journey. He clamored for the window, hooted at passengers in passing cars, and nearly killed us with his attention-getting antics. But even his patience deteriorated when he realized that he had to be confined to Jerry's arms for the entire trip.

We arrived somewhere around eleven thirty that night, looking as if we'd spent the last hour and a half with Torquemada. We were beat. My eyes were red from crying; Jerry's voice was hoarse from shouting. My mother was certain we'd been in an accident, or at the very least just barely avoided one, and kept pressing us for the truth. "You can tell me," she assured us, as if we were concealing a hit and run. Surely nothing less could have left us in such a state. We gave her a brief rundown of the trip and had only anecdotally reached the Lincoln Tunnel when she called it sufficient. She advised a good night's sleep for all of us.

Boris had other ideas. He had a second wind. He had never been in another house and he was entranced. He wanted to touch everything, but unfortunately everything wasn't touchable. Neither Jerry nor I had anticipated the fragility of my mother's home. Likewise, my mother had not anticipated the dexterity of an ape when she'd invited us. It was too late for second thoughts. A vase that had come through the Revolution unscathed didn't survive fifteen minutes with Boris. We left my mother sweeping up the pieces and carried Boo upstairs to figure out how to put him to bed.

Ape nests are simply constructed. There is no elaborate weaving or knotting. The animal sits or stands on a suitable site and then

> reaches out in one direction after another, grabbing nearby branches and bending them in towards himself. . . . These actions form a basic platform. The nest may then be improved by pulling into position smaller twigs and foliage, tucking them under or around the body.
> *Men and Apes*

The first thing we had to do was devise something that would be comfortable for Boris and safe for the house. Simply to lock him into a bedroom would jeopardize both. My mother's decorating taste ran to the warm cluttered look, and there was no way to strip a room down to its bear unbreakable essentials in a single night. We wracked our brains, rummaged, and finally found an old crib that looked as if it could serve as a cage if we covered it with a weighted piece of plywood. In fact, we were quite pleased with ourselves when we snuggled Boris in and slapped on the improvised lid, securing it with a carton of books. We were not quite as pleased ten minutes later when, after much screeching and hooting, there was a loud crash and Boris came running out into the hall. The plywood was not going to work. Boris had pushed it off, books and all, and would undoubtedly be able to repeat the feat at his pleasure. We were becoming a little desperate. The trip had sucked our humor dry. It was late and we were tired and Boris was raring, willing, and able to go on . . . and on.

"Aha!" Jerry said.

"Aha yourself," I snapped, rediapering Boris, who'd gotten very excited and very wet.

"We'll turn the crib upside down. Put the mattress on the floor, take out the spring. It'll be perfect."

It sounded good. Anything would have sounded good. Had Jerry told me we were going to tie Boris spread-eagled to the bed, I would have said it sounded good at that point.

"Come on and help me," Jerry said.

"I knew there was a hitch."

"How do you expect me to turn it over by myself?"

"How do you expect me to help you and watch Boris at the same time?"

"It'll only take a minute."

"How long did it take him to break that vase?"

"There are no vases here. Come on."

I looked around—chair, bed, pillows, bureau, mirror, table, pictures, lamp—they all looked sturdy enough to withstand sixty seconds of unsupervised Boris. "Okay. Let's go."

We inverted the crib in forty-five seconds, just about the same time it took Boris to step innocently on the night table and invert the lamp. Fortunately, only the bulb and my nerves shattered. We tucked Boris under the crib and congratulated each other. We thought we were geniuses; Boris thought we were Gestapo. For once we were too exhausted to feel guilty.

The next day Boris had his first taste of the real outdoors: unadulterated country; and he went ape. Although he stuck pretty close to my side, he pretended I wasn't there. Losing himself in fantasies of freedom, he tasted the grass, scampered around the bases of trees, and even nibbled insects. He'd wet his finger and then scrunch down on an ant, examine the remains for a few seconds, then pop them into his mouth. The process seemed as enjoyable as the taste, and by the end of the afternoon he'd devoured an easy dozen armies.

My brother and sister-in-law, their daughters and dog arrived the following day and the weekend took on circus proportions. Their dog, Happy, was a small wire-haired terrier. We realized at once that he wouldn't stand a chance against our four-legged terror, so we confined Ahab to quarters. But Boris seemed a fair match; at least their teeth looked about even. The game of the day was a sort of touch tackle. Boris would chase Happy, tease him into attacking, then leap out of the bite line onto a table. Every so often he would stand his ground and Happy would mouth his arm. Boris in turn would mouth Happy's ear. They didn't bite each other, which was rather remarkable, and they appeared to enjoy the roughhousing.

My eight-year-old niece had brought along a large ball, the kind that you sat and bounced on. Boris watched the children use it for a while and then took over. He clung to the handle, rode it like a rodeo star, bounced it with his feet. His performance delighted everyone except my niece, who did not appreciate his appropriating her toy and was definitely unhappy with his finale. Just as I was about to give the ball back to her, Boris sank his teeth into the rubber. I told her that I was sure a tire patch could fix it up in a minute, but she didn't believe me. She gave Boris the dirtiest look an irate eight-year-old girl can muster, refused to have anything to do with him, and referred to him as "that mean monkey" for the remainder of the weekend. Needless to say, it put a strain on convivial interaction between my brother's family and ours, a relationship that was already handicapped by Ahab's unquenchable desire to eat my brother's dog.

Inconceivable as it was, the ride back to the city was worse than the ride out. Shep was determined to bring home the remains of a gargantuan outdated stereo system, which left very little room in the back seat for Ahab, and Boris was determined to drive. When we weren't swatting Boris's hands off the wheel or passing paper towels to the back seat, Jer and I were being assaulted by an intense verbal description of every component part in the stereo as well as in Ahab's upheavals. It was a madhouse on wheels. The capper came when a nice toll taker at the Lincoln Tunnel reached in to pet Boris and almost lost her hand to Ahab. When we got home, we all looked as if we needed a vacation in the country.

CHAPTER 9

* *Boris, We're Going to Make You a Star* *

At fourteen months, Boris blossomed into a remarkably beautiful animal. His scraggly, porcupinelike baby hair had grown long and silky, and his creamy, plump cheeks had grown plumper and creamier. His range of facial expressions was incredible. Although Jane van Lawick-Goodall had categorized only five key chimp expressions, with two variations, I'd seen Boris run through almost twice as many in a day. Where the Goodall chimps had only two expressions for pant-hoots—the series of breathy "hoo" sounds chimps make when they eat, cross valleys, join other groups, or call friends—Boris had four or five. He could pant-hoot out of the corner of his mouth, with lips pursed or straight, or while chewing a banana. He enjoyed having fun with his face.

The little narcissist loved gazing into his own nut-brown eyes.

Photo: David Sagarin

He had the most fun with his face in front of the mirror. It kept him more amused than any of his toys. Nothing enthralled him more than the sight of his own lips puckering at himself from the glass. He would perform tricks for himself, jump up and down, and peer at his own image from between his legs. There was no part of that little narcissist that he didn't find fascinating.

We began to wonder whether we had nipped his career in the bud. He certainly had the ego requirements for a celebrity. Watching him delight himself at the mirror one evening, it struck me that we had probably expected too much of him much too soon. He was ripe for some sort of stardom. Perhaps not television or films, I wasn't up to that again, but . . . I had it, or at least thought I did. I smacked my hands together and jumped to my feet.

"Mosquito?" Jerry asked.

"Nope. Brilliant idea."

Jerry's look indicated that he would have preferred the mosquito. "Shoot," he said, the prisoner telling the firing squad.

"We'll do a photographic children's book starring Boris. It'll be sensational."

Jerry "hmm'd," which gave me a chance to shower him with my enthusiasm. By the time I had finished I was convinced the book would sell like *The Little Prince* and *Bambi* combined, and Jerry had conceded that it might work.

While my enthusiasm was still in the infectious stage, I rang up David Sagarin. "It'll be a snap," I told him. "You do the pictures, I'll write the text, we'll split the proceeds fifty-fifty, and we'll all take a holiday in Bermuda."

Bermuda sounded good to David.

"Wonderful," I said. "We can shoot the photos this weekend."

"Fine with me, but what's the book about?"

"I'm . . . not sure." Which was to say I hadn't even thought of it.

"How can I take photographs for a book without—"

"A birthday party," I said suddenly. It was, I thought, an inspiration. It was, in fact, the first thing that came to mind. "Boris will be the present."

"Yeah . . ." David said slowly. I could see his mind positioning shots through balloons, wide-angling effects through party favors. David was a very creative photographer.

"We'll title it *A Birthday Present Called Boris,*" I said. "I'll buy the props and we can probably shoot the whole thing in an afternoon."

"Sure," David said.

Which just went to show how much either of us really knew about Boris.

Boris relaxed and played normally all that week, oblivious of his impending rise to fame.

I, on the other hand, spent three hectic lunch hours garnering the most colorful birthday paraphernalia on the market. I began decorating the apartment early Saturday morning, and by the time David arrived at noon, the place looked like Forty-second Street after Lindbergh had crossed the Atlantic. Streamers wiggled out of our bookcase, crossed the room, and joined pole lamp to chandelier to door frame. Cups, hats, candy, and horns festooned the table, and bowls of pretzels and potato chips were all about. Boris had watched the proceedings with a very interested expression, but I'd kept him confined to his cage to make sure everything would be just right for the shooting session.

"Don't you think you might have . . . overdone it?" David asked as he began setting up his lights, trying to keep them clear of the streamers.

"I thought a colorful background would be nice." I didn't dare look at Jerry, who'd told me at nine thirty that morning that I'd overdone it.

"Let's cut some of those things, remove a few hats, and—" David glanced around. "You've enough potato chips for an army. How many kids are coming?"

"Kids?"

"It's supposed to be a party, isn't it?"

I felt sick. I'd become so embroiled in the set, I'd forgotten the cast. I'd even sent Shep out to visit a friend so he wouldn't get in the way. "I thought we could . . . could . . . er . . . get them at another shooting," I said.

"Just as well," David said. "We'll see what we can come up with today and then we'll know what we need." David was nice.

While David arranged his equipment, I dressed Boris in his red and white polo shirt and brushed his hair. He was unnervingly subdued—all eyes, all innocence, just watching the preparations. While I held him, he munched a banana, like a gunfighter chewing a wad before the showdown. I didn't like it.

"Why don't you let him walk around and warm up a bit?" David suggested. An even dozen famous last words.

I took Boris from my lap and sat him on the couch. He pushed the remaining banana into his mouth, tossed the peel into the air, and he was off!

Before David could blink a shutter, Boris hopped into the box of photo equipment and began casually gnawing on Pentax lenses. David grabbed for him, but Boris was out of reach and scaling the lights before anyone could stop him. By the time I managed to scoop him into my arms, David's carefully positioned lights were askew, there were wires strewn all over the floor, and Ahab was happily finishing off a bowl of spilled potato chips.

"I think we need some direction," David said. He suggested we put Boris on the table and have him eat some marshmallows for openers. What could be easier? Boris adored marshmallows. I sat him on the table and pointed to the bowl, then stepped back.

It was a mistake.

Boris didn't reach for the marshmallows, the hats, the horns, or any of the things that would have made a cute picture. Instead, he bent over and took the tablecloth in his teeth and began to chew it up.

"No!" I shrieked, which sent Boris racing across the table and back again, leaving a wake of toppled paper cups, hats, horns, and marshmallows. Within seconds, he'd pulled down the streamers I'd so carefully crisscrossed and was leaping up for the chandelier with a mouth full of candy. Jerry caught him in mid-jump, and Boris hooted his displeasure.

"I . . . I think I got a few shots," David said.

I hoped so; retakes were out. The table was a wreck. There was only one thing to do—change the book concept. We'd have Boris *really* run amuck at the party. I thought this was a stroke of genius; David refrained from committing himself.

"How about a shot of Boris running down the hall trailing a crepe paper streamer?" I suggested.

"How are you going to get him to run down the hall?" David asked. He was becoming increasingly suspicious of my suggestions, I noticed.

"Simple. I'll hold him at one end of the hall. Jerry will go to the bedroom at the other end and hoot."

"Jerry will . . . hoot," David repeated slowly.

"Boris always comes when we hoot," I explained.

David nodded. It was the sort of nod you give people who tell you that Martians are infiltrating the government. I ignored it.

"Wait a minute," Jerry said. "What about Ahab?"

We could keep Ahab out of the living room, but the hall was his turf, and when something belonged to Ahab, it belonged to Ahab. We decided to lock him in the bathroom. David felt this was a wise decision.

David rearranged his lights while I recombed Boris's hair and Jerry disappeared into the bedroom.

"Tell me when!" Jerry shouted.

I handed Boris the streamer. "Now!"

David raised his camera. Jerry began to hoot. Boris started to run—but not down the hall. With split-second timing, he about-faced and tore into the living room.

I started to hoot to get him back into the hall. Jerry, unaware of what had happened, hooted louder. Boris tuned us out and jumped back into the box of lenses. David hooted something obscene and raced in to protect his equipment and I followed. Jerry was still calling Boris's name and hooting from the bedroom.

"Boris is in the living room," I shouted.

"What's he doing there?" Jerry yelled.

"Everything. Help!"

When Jerry entered, Boris was standing on the couch holding one of David's very expensive lenses—precariously. David looked a little pale.

"If we try to grab him, he'll toss the lens the way he did the banana peel," I explained, which did absolutely nothing for David's color. I told him I was going to try hooting softly.

David expressed his confidence in the hoot recall by closing his eyes.

I moved to the edge of the couch and did my best imitation of Boris's own soft, breathy, "hoo-hoo-hoo." There was a half-moment of indecision, then Boris laid the lens down gently and scampered to me.

"See?" I said proudly, petting Boris.

"Yeah," David said, petting his lens.

It was one of those poignant moments.

We tried it again, this time with David's equipment carefully locked in another room, but Boris was determined not to run down the hall with the streamers. He'd either take off with the crepe paper or he'd get halfway and then run back into my arms before David could shoot. By six o'clock, we were whipped.

"I guess we should call it a day," I said.

David said he had another name for it. We all agreed we had

been overly ambitious to think we could wrap up the book in one session.

"Next time, I'll let him run off some energy before the shooting," I promised. David suggested a week. We left the date for our next get-together open.

* *The Hairy Genius* *

Man excepted, probably no animal has been given more intelligence tests than the ape. And, man included, probably no other animal has done as well. Chimps have shown that they can operate slot machines, do as well as an average second-story man in opening locked cabinets, and match a goodly portion of aspiring artists on canvas, to say nothing of the fact that they've already proved they can man space capsules. That chimps seem unremarkable in their wild state, by human standards, is mainly due to fear. One whomp! by a chimp mother is enough to inhibit even the most primal of a youngster's investigatory urges. But a chimp raised by humans is an entirely new batch of coconuts.

> Everything around it is probed and prodded, opened, turned, twisted and bitten. Large structures are clambered over, shaken, slid down, swung around, and jumped on. Small objects are tasted, smelt, touched, carried, torn or squashed. Anything new or unfamiliar receives immediate attention and is tested and investigated until it has become completely familiar. In this way the environment gradually becomes known and experiences mount up. *They are stored away in the memory to be utilized in an emergency when a new reaction has to be produced* (Italics added).
>
> *Men and Apes*

After the photographic session with David, Boris's exploratory forays mushroomed. It was as if he'd recognized his own limitless potential in our shamelessly adoring eyes. He began to examine things with an earnestness that transcended curiosity. Where he had simply used to watch us do something, he now studied it. When I'd turn on the television, he'd press his nose right to my finger while I pulled out the *on* switch. Thinking it cute, and being always in favor of higher learning, I showed him how the button made the set go off. That adorable thought cost us fifteen dollars. Two days after he'd watched me do it, Boris tried an on-off strobe effect with

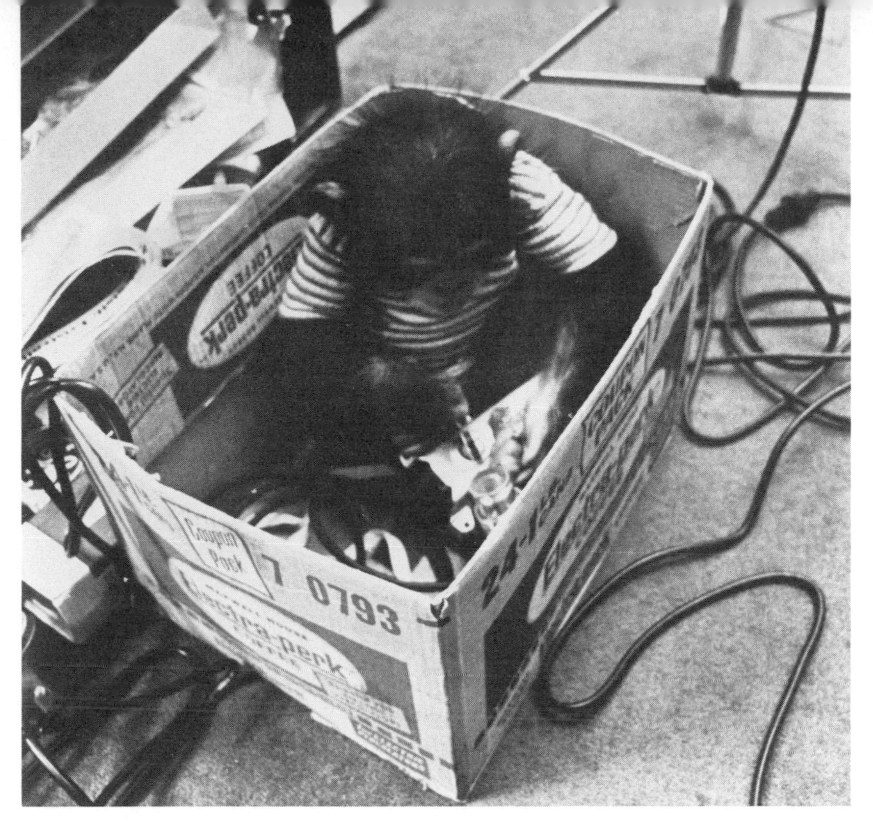

Exploring whatever was off-limits gave Boris his finest hours.

the TV and pulled the button out of its socket. So much for being in favor of higher learning.

But the television was the least of our concerns regarding Boris's rapidly growing intelligence. He had discovered the hiding place for his cage key, and could open the lock! Right from the start, Boris had watched our opening and closing of his cage with more than passing interest, but we never paid much attention to it. Little did we suspect that it was anything more than chimpy curiosity. But one afternoon, as Boris was nondestructively ambling around the apartment, lulling me into a false security, I turned around and found him standing on the secretary and opening the drawer where we kept his key. Now the secretary had four identical drawers, and the drawer where his key was kept had some ten different keys in it. At first I thought it a coincidence that he had happened to open that particular drawer, but it was no such thing. He put in his finger, pushed the various loose keys around, and extracted the one small key that fit his lock. I was stunned. He was only fourteen months old!

As I drew closer, he regarded me uncertainly and slightly impishly. Then he pretended to study the key as if he'd never seen an object like it before. It was an obvious pretense, because he kept looking up at me and then quickly hunching back over the key. He was waiting for me to move away. I stepped back, just far enough to remain in reaching distance should he decide to put the key to his standard gullet test, but he didn't. He jumped from the desk and went straight to his lock. Within five minutes, he'd inserted the key.

My immediate reaction was overwhelming awe. (I have a proclivity for hyperbolic emotion.) But then I realized the possible consequences of Boris's amazing feat, and a disquieting chill scaled my spine. Although it was unlikely that Boo could ever hide the key in his cage without our knowledge, it was not an impossibility. Many a night, I'd put Boris back in his cage still clutching a spoon or a toy. What was to stop him from palming the key? He could escape in the night and perform unnatural acts upon our stereo. But worse, and more likely, he could lock himself in, which would mean calling a locksmith, or buying a hacksaw, or chopping again into our fragile finances somehow or other. He "hoo-hoo"d softly as he placed and replaced the key in the lock; the joy of discovery. Could I be so cruel as to quash such inventiveness, not support such achievement? Were material things like our stereo, books, lamps, paintings, and china worth the risk of jeopardizing Boris's thirst for knowledge? My soliloquizing quandary made Hamlet's seem like a decision between vanilla and chocolate. I'd progressed to debating the issue with myself on a metaphysical level when I looked down and saw that Boris had dropped the key and was now quietly sucking my typewriter.

"Boris might know how to open his lock," Jerry said when I told him what had happened, "but he doesn't associate it with escape."

"I don't know about that," I protested. "He might not be thinking of escape per se, but he knows that somehow the key is related to his being in the cage. Maybe I should have scolded him."

"Ridiculous."

Two days later, Jerry had to buy a hacksaw and a new lock. He had the store make up two extra keys, both of which I promptly hid in the bottom of my lingerie drawer. As long as Boris didn't turn out to be an underwear freak, we were okay.

* *Time Out* *

There comes a time in every pet owner's life when it becomes necessary to leave the fluffy, scaly, or feathery dependent in someone else's care and take a vacation. This time is often heralded by the owner's fatigue, a tendency toward short temper, and a definite foreboding of imminent mental collapse. Our time arrived at the end of August.

Jerry and I needed a break, desperately. Our lives had revolved around Boo-Boo so intensely for months that we no longer knew what it was like to enjoy things outside our apartment without working them in and around Boris's schedule. Unlike having a child, owning a chimp required a unique sort of attendance. Spontaneous entertainments were out. Whereas we could always drop Shep off at a friend's house, Boris's acquaintances were limited. And of those who might be willing to chimp sit, none were actually capable. Boris was growing stronger daily, and diapering a recalcitrant ape was like trying to pin a daisy on the Minotaur.

But we had to get away. I was bursting into tears every time I lit the wrong end of a cigarette, and Jerry was erupting like Vesuvius at the clang of his typewriter's carriage return. Quiet, remote Post Lake, Wisconsin, where Jerry had fished and gamboled in sensuous solitude as a boy, was where we wanted to go. But what were we to do with Boris?

As far as boarding was concerned, Ahab was no problem. We had long ago found a gutsy kennel on the East Side that could not only handle him, but liked him. They even took him for walks with other dogs, which never ceased to amaze me. The only explanation we could come up with was that when he wasn't protecting us, Ahab relaxed into normal canine behavior. The kennel had informed us, though, that Ahab usually became much tougher toward the end of his stay, because he became protective of the handler walking him. Ahab, like Cerberus, enjoyed responsibility.

There was really only one place to board Boris, and that was Trefflich's. We knew that he'd be taken care of there, that they had the facilities and knew his diet, but it made me feel as if we were sending Oliver back to the orphanage. Would Boris understand that we hadn't abandoned him, returned him to sender, so to speak?

"He'll understand when we come back," Jerry said, pulling his fishing equipment from the closet.

I was holding Boo on my lap, stroking his hair while he planted little heart-wrenching kisses on my neck. "But he's so little, so vulnerable. He'll be lonely."

"They'll play with him. There'll be other animals around."

"He'll miss watching television."

"It'll be good for him." Jerry began to string new line on his casting reel. "He watches too much anyway."

Boris reached over my neck, grabbed hold of the fishing line, and yanked. The casting reel and tackle spun to the floor and rolled into tangles. Boris was after it like a shot.

"Stop him!" Jerry shouted.

As a mother I'd learned to move fast, but I was no match for Boris's jungle swiftness. Boo had Jerry's line looking like a macramé belt, albeit a poor one, in record time. The decision had been made. Oliver was going back to the orphanage for a week.

Boris hooted fiercely when we brought him into Trefflich's shop. Whether it was excitement or outrage we weren't sure, and neither Jerry nor I wanted to speculate. Henry Trefflich was gregarious and friendly as usual, and told us we hadn't a thing to worry about, that Boo would have plenty of company and fun. To prove it, he brought us into the back of the shop where two chimps, older than Boris, were reaching out through the screens of their tiny cages like jailed winos asking for a drink. Guilt seized my heart. The cages were one-tenth the size of Boris's. This was Henry's idea of fun?

"This one," Henry said, pointing to the mottle-faced chimp on the left, "plays the piano. And Tina here walks a tightrope." He took Tina's hand and gave it a little shake. Tina barked like a seal. Anthropoidal considerations aside, she was the frowziest aerialist I'd ever seen. Her hair was sparse and matted and her face a discordance of standard ape features, with nostrils too flared and ears too pointed. Also, her front teeth had been pulled, but I guessed that was show business.

I looked at Boris and realized, perhaps for the first time, how very handsome he was. Aside from aesthetics, he had an engaging innate innocence that was totally lacking in the other two. I suppose carny life could have hardened them, but I chose to think that Boris was just born special.

Meanwhile, Boris was gouging into my bare shoulders with his nails. He had hooted loudly when he'd first seen the other chimps, but a prescience of an as-yet-unknown but definitely disagreeable thing about to happen had subdued him. He had no intention of letting go of me. And I, after viewing those cages, was reluctant to part with him.

Jerry sensed this with my first backward step. "We're going to Wisconsin," he said, putting a firm hand on my arm. "Boris will be fine."

"He'll get plenty of play," Henry assured me, and before I could protest, Boris was wrenched from me and locked in a cage that would have been close for a squirrel. He pushed his arm through the screening and I had to look away.

I pulled out Boris's security blanket, my old sweatshirt, and handed it back to Jerry. "I brought this to leave with him."

Without seeing Jerry's face, I knew my maternal thoughtfulness had embarrassed him in front of Henry, who obviously scorned such tender foolishness. I also knew that Jerry couldn't bear the guilt of denying Boo his favorite sweatshirt while we were off frolicking with nature in Wisconsin. He cleared his throat and did a hushed man-to-man with Henry, something about how it would "make the wife feel better."

"Sure," Henry said. He opened the cage and I turned in time to see Boris clutch the sweatshirt to his chest. I felt better and worse at the same time, if that's possible. I was relieved that Boris had the shirt to comfort him, but the way he clutched it convinced me that we were doing the most heinous thing since the slaughter of the innocents.

When I said good-bye to him and started to walk away, he popped his forefinger in his mouth and "hoo-hoo"d plaintively. I was a mess for the rest of the day.

CHAPTER 10

* *Home and Running* *

Our vacation was everything it was supposed to be—restful, rustic, animal-free, and fun. But we'd missed Boris, even Shep had, and were anxious to reclaim him as soon as possible. As we were unpacking our suitcases, Jerry volunteered to go and get him. I raced out to the supermarket for a welcome-home bunch of bananas while Jer taxied downtown. I'd only been back in the apartment an hour when they returned.

Boris literally jumped into my arms, hammering out "hoo-hoo"s with pneumatic drill rapidity. Jerry looked bushed but happy.

"I'm glad to see him like that," he said.

"Didn't you expect him to be excited to see us again?" I asked.

"That's what threw me. When I opened his cage down at Trefflich's, he refused to come out. He took one look at me and turned his back."

"Boris? Are you sure he recognized you?"

"He recognized me, all right. He was just ticked off about having been left there, and it was his way of getting back at me."

"Boris is not vengeful."

"How would you greet me if I'd locked you away somewhere for a week?"

"Hmmm. You're lucky all he did was turn his back."

"I believe it. Henry said that animals are wary of new surroundings, and sometimes, once they've gotten used to a cage, they're as frightened of coming out as they were of going in. But he also told me that he'd had Boris out of the cage quite a bit. Boris was just angry at me and showing it."

I tightened my hug on Boo and he tightened his on me. "You're home now, mush face," I murmured. "Safe and sound." I put him down and went to get a banana. He followed me into the kitchen. He wasn't taking any chances.

When he saw the bunch on the table, he didn't even give himself time for a food bark. He scaled the kitchen chair and leaned over to start biting.

"Whoa! They must have starved him," I said. I tore off a banana and handed it to him. He downed it as if he was loading a clip into a pistol, putting the tip into his mouth and then jamming the rest in with his hand.

"Doesn't look like it to me," Jerry said, patting Boris's ample belly.

"You're right," I said, surprised. "He's actually chubby."

Boris reached for another banana.

"He sure is hungry, though." I picked him up and rubbed my hand over his stomach. It was as hard and as round as a basketball. I didn't like it.

"So we have a pudgy chimp," Jerry said. "What's wrong with that?"

"Nothing—I guess." But I was disturbed. The underside of an ape's belly should not be as hard as the topside of a tortoise's shell, and Boris's was.

"He's probably into a growing stint," Jerry assured me. "Look, he wants another banana."

He not only wanted another banana, but another after that, and then some plums, lettuce, raisin bread with cream cheese, and then some marshmallows. His whole focus had swung to food, and his appetite was nothing less than obscene. We had never seen him eat with such gusto, such relish, such savage determination. It was almost frightening.

We finally distracted him with some toys from what was taking on overtones of obsession. He was pleased to rediscover the toys and spent the next half-hour banging his play lawnmower up and down the hall, happily hassling Shep, who immediately accused us of playing favorites by reclaiming Boris and not Ahab.

"I'll get Ahab tomorrow," Jerry said. "I'm beat."

"You weren't too tired to get Boo-Boo," Shep said sardonically. The kid was a psychic marksman. I could feel Jerry lurch with the hit. I handed him a roll of paper towels and a bag. An hour later our household was back to its normal chaos.

By bedtime, we were all exhausted. But as I laid Boris down on the couch to put on his nighttime diaper, I noticed that he was breathing strangely, heavily. It sounded as if he was doing soft versions of his excitation hoots. Then I saw his stomach. It was terrifyingly distended and rock hard. True to form, I panicked.

"My God, Jerry," I wailed. "What's wrong with him?" I took him in my arms and began pacing the floor, patting him on the back, telling him it (whatever "it" was) was going to be all right, trying to block out the awful, rasping sound that was emanating from his lungs.

"He doesn't look sick," Jerry said. "No fever."

"Listen to him! Touch his stomach!" I paced faster, trying to run from the rapidly encroaching fear that Boris was about to explode. Boris, oblivious to, though not unaware of, our concern, was casually reaching out for lamps, books, whatever was in striking distance from my shoulder as we circled the room.

I felt tears well in my eyes, but I was too frightened to cry. Boris's breathing had taken on the sound of heavy machinery in need of repair. He was virtually chugging.

And then, all of a sudden, it happened. There was a low rumbling in Boris's chest and a Krakatoan burp erupted, followed moments later by another.

"It's gas!" I shouted, with such exuberant relief that Shep came stumbling out of bed, convinced the building was in imminent danger.

"Boris has gas," I explained.

From beneath half-closed lids Shep gave me a distinctly dirty look and shuffled back to his room, muttering something about false alarms being punishable by law.

Boris, unconcerned by our relief and Shep's distress, managed to decapitate a ceramic donkey by knocking it to the floor, and he burped on. In less than an hour's time, his stomach had softened and his breathing had returned to normal. Our guess was that his diet of chimp food and Trefflich's and his urge to compensate for it upon his arrival home had been responsible. Probably nerves, too.

* *Reverberations* *

Our week in Wisconsin and Boris's at Trefflich's were to bring about more consequences than we could have imagined. Although Boris was truly happy to be home, as evidenced by his appetite and excited play with toys he had previously taken for granted or ignored, he was not at all pleased to be back in diapers and clothes. His au naturel stint at Trefflich's had convinced him that his jungle birthday suit was all any self-respecting ape should wear. In theory, Jerry and I went along with this, but in reality it wasn't possible. The heat in our apartment came and went at the whim and erratic sleep habits of our superintendent. Boris's proclivity for colds and our fear of pneumonia necessitated his wearing shirts. As far as diapers went, well, it was our dining room, after all.

But having gone sans Pampers for a week, Boris was determined to assert his new liberation. He didn't take to burning them, but he did do a neat job of shredding them into little pieces and hurling them all over the dining room.

I had taken to putting training pants over his diaper, but he'd learned to remove these too. And, for more than appearance, I was concerned. These were the days when Pampers required diaper pins, and knowing Boris's abundant oral tendencies, I feared the worst. Sure enough, it happened.

I entered the dining room just as Boris began to shred his twelfth diaper that day, then noticed the diaper pin in his hand. I shouted "No!" and hurried to get him out of the cage, but by the time I had him in my arms, the diaper pin was gone.

It had happened so fast, I was sure it couldn't have happened. I searched the bottom of the cage, but no diaper pin. Boris observed my search with delight and joined in, tossing his blanket in the air and flipping over his bed, crawling under the buffet alongside his cage. But no diaper pin. In desperation, I turned Boris upside down and shook him like a piggy bank. Boris gave a few hoots, but nothing more.

I was distraught; Boris was ebullient. He kept climbing on my lap and dangling upside down. I kept flaying myself with the thought that he'd swallowed the pin. Finally, Jerry and Boris himself convinced me that I had to be mistaken. Boo's appetite was fine, his spirits better. Obviously, the pin had fallen behind something outside the cage and would soon be found; obviously. But it didn't do a thing for my sleep that night.

Boris was extremely playful the next morning and, to my surprise, much more amenable to diapering. He didn't try his back-to-nature routine once that morning, and I went to work feeling secure, happy, and slightly foolish about my diaper pin terror. By the following day, I'd forgotten it completely. At four thirty that afternoon, Jerry phoned to say that Boris had uneventfully (barring a few screeches from Boo and some mild hysteria from Thelma) passed an inch-and-a-half closed diaper pin.

That evening I invented an adhesive tape diaper tab, which a year later became standard on disposable diapers. I received neither monetary recompense nor recognition for my handcrafted safety device, just wonderful peace of mind. It was one of those inventions definitely born of necessity.

* *Physical Fitness and Photos* *

For many weeks after Boris's stint in what I'd come to call Trefflich's Tiger Cage, Boo would spend large portions of the day working out and regaining his muscle tone. It was his own self-devised regimen of headstands, somersaults, hand and foot swings, each repeated over and over again until he was either pleased with the result or tired of the effort.

Outside the cage he did a swell imitation of a short-legged long-distance runner. He'd wait until Ahab was about six muzzle lengths away and grab one of his bones. Then he'd "hoo-hoo," attract Ahab's attention, and they'd be off on a chase around the apartment, which would end only when Boo tossed away the bone to avoid Ahab's precariously close jaws. Should Ahab leave the bone soon after retrieving it, Boo would snatch it again and the race would be repeated. Bone relays, as we came to call them, were also very similar to ice-cube hockey, but much more dangerous for the spectator and the field of play. Once, Boris tossed the bone up on the kitchen table,

A classic example of Boris brinkmanship.

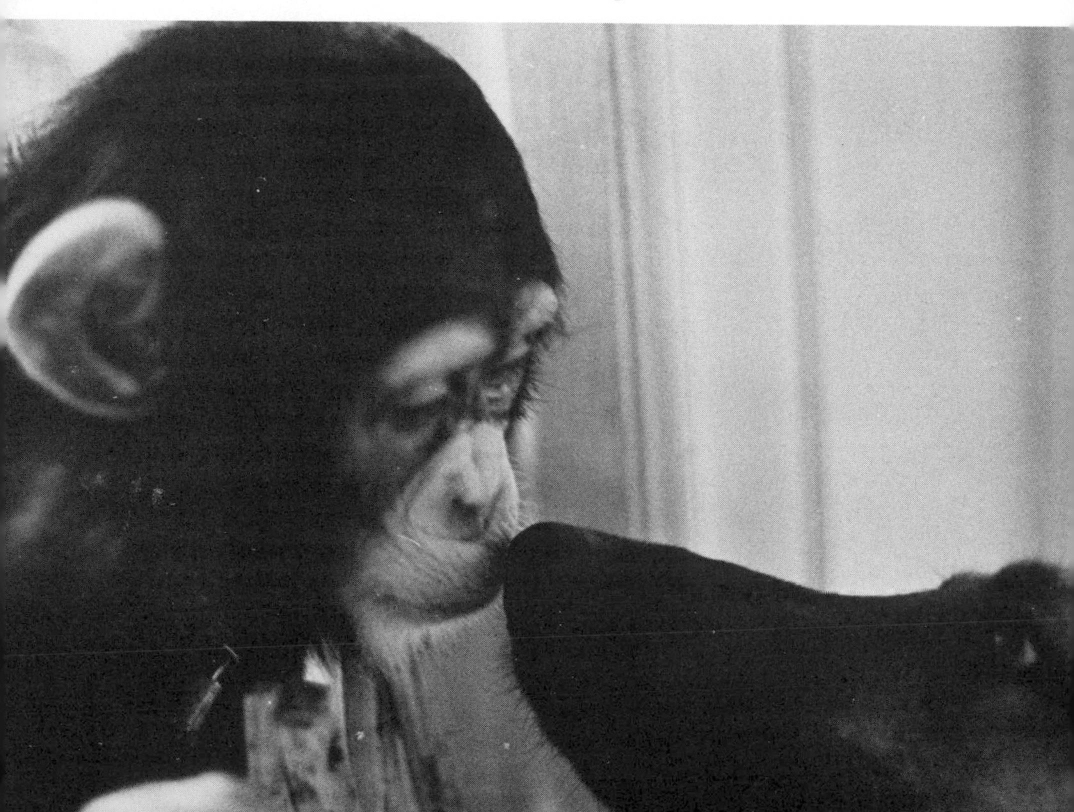

where I had carefully placed six beautiful long-stemmed crystal goblets to dry. Before anyone could stop him, Ahab's front paws were on the table, his jaws on the marrow, my goblets on the floor. Jerry was impressed with my creative variations on gutter expletives.

Boo's tone-ups put him into grand emotional and physical shape, and we decided to try another shooting with David Sagarin for the children's book. This time, David thought, we could work in a studio and get some good shots of Boris doing some cute but controllable things, like eating and jumping out of the gift box that, according to our tentative story line, he was supposed to arrive in.

The studio was a loft on lower Park Avenue. It belonged to a vegetarian photographer friend of David's and was equipped with as many health food gadgets as photographic implements. There was a machine for making yogurt, another for turning carrots and cucumbers into juice, and still another for whipping that juice with wheat germ and Tiger's Milk into a frothy energy cocktail. The guy also had machines for making noodles, shredding coconut, and fermenting wine. When we arrived, he introduced himself as Oscar and told us that it meant *Pure*. He was smoking a home-grown cigarette.

Oscar the Pure mixed himself a "carrotini" while David unrolled a large cylinder of no-seam for a backdrop. Boris was very quiet, which was unusual—and ominous.

"Why don't you let the little guy look around?" Oscar suggested.

"Well . . . I'm afraid he might break something," I said honestly.

"Ah, he'll be all right. Let him have a little fun." He held out his arms. "I love monkeys."

"He's an ape," I said.

"Apes, monkeys, same thing. I love 'em all." He clapped his hands. "Come on, Butch."

"Boris."

"Come on, Boris."

I looked at Jerry. He gave me one of those you-heard-him shrugs. I looked at Boris. He gave me one of those gee-gosh-golly-mom-would-I-do-something-wrong? expressions. I closed my eyes, crossed my fingers, and put Boo down on the floor.

He went straight to Oscar the Pure and sniffed his feet, which apparently struck Oscar as the funniest thing he'd ever seen, because he broke into explosive laughter and began slapping his knee.

Boris jumped back. His hair bristled and he began to "hoo-hoo" defensively. I was just about to explain to Oscar that he'd frightened Boris, that it was best to come on gently, that he must not try to grab him, when he "hoo-hoo"d back at Boris and tried to pick him up.

Boris shrieked and tore like hairy lightning across the loft. In seconds he was scaling the steam pipe, which unfortunately connected with a grand network of ceiling fixtures, and Jerry and I spent the next half-hour talking him down. Oscar the Pure got bored with our efforts after the first fifteen minutes and told David he'd be back around five.

Once we had Boo back in our clutches, I was reluctant to let him go again. The ceiling in the loft was a healthy fifteen feet high, and Boris kept eyeing it, sneaking little upward looks like a shy Icarus. I pretended not to notice and kept his hand locked firmly in mine.

David, who'd been ready for a while, suggested we start with something simple. All we needed was Boris standing in the large, gaily wrapped box, looking as if he'd just happily popped up. Simple.

David put the box on the no-seam and adjusted his lights.

"Okay. Put him in," he called.

Boris loved boxes, jumped into cartons all the time, and rode around in them in the apartment. They were among his favorite toys. I naturally assumed that as far as Boo was concerned all cartons were alike. I have since learned that natural assumptions, like mushrooms, should be used with caution. Boris freaked.

Not only wouldn't he get into the box, he wanted out of the studio that contained it. His body stiffened every time I tried to lower him, and he clung so fiercely to my shirt that his nails went through the fabric.

"I thought he liked boxes," David said.

"He does. Doesn't he, Jerry?"

"Loves them."

David just nodded.

"Boo-Boo," I said quietly, "under this wrapping paper is a box. You'll love it once you're in it." And with that I plunged him into the carton and hurriedly stepped back. David managed to get three or four shots before Boris's plaintive, terrified screeches got to me.

"How do you think they'll look?" I asked, calming Boo with a banana.

"I don't know," David said. "They might work. I'll have the contacts by Wednesday."

On Wednesday night I discovered what a chimp in the throes of stark terror looks like. His upper and lower teeth are showing, his jaws are open, and he looks as if he's grinning.

"It'll work in the book," David said, "as long as no one who reads it knows anything about chimps."

The chances of getting a terrific photo of Boris in the box were equal to the likelihood of Jane van Lawick-Goodall's reading the book. We decided to run with it.

* *Chips and the Chimp* *

On Wednesday nights Jerry played poker. The game was usually played at one or another of the guys' apartments, though it had never been at ours. After six months, Jerry felt he had to offer our home for the game.

"Did you warn them about Ahab and Boris?" I asked when he told me.

"I didn't warn them," he said. "They knew about Boris, and I just mentioned in passing that we had a dog."

"Ahab is not your ordinary dog."

"I never said he was ordinary."

I could see that Jerry was getting defensive, so I dropped the matter and let my anxiety eat me quietly. By Wednesday night I envisioned my stomach lining as Swiss cheese.

At seven thirty the guys began to arrive, and the entire building knew it. Ahab tightened up with each successive bell ring. By the time the last player came to the door, Ahab was a four-legged hair trigger.

The infamous Birthday Book photo, or—why isn't this chimp laughing?

Photo: David Sagarin

The game was to be played on our dining room table, to Boris's delight and the men's increasing unease. Boris's cage was a scant three feet from the table. I noticed that those who entered the room first, who joked a bit about dealing Boo in, chose chairs farthest away from the cage.

The pregame conversation was basically a monologue by Jerry, who had launched into his standard Ahab-defense lecture, explaining the difference between vicious and assertive dogs, telling the men that what most people call a vicious dog is merely a bold animal that's been poorly handled, avoiding the fact that we, on the other hand, did have a vicious dog.

Boris clung to his chain link, listening as intently as the men, and "hoo-hoo"d whenever they broke into nervous laughter, which was often.

Ahab positioned himself at the dining room entrance and stayed there, poised.

The game began quietly, and after a few hands the men began to relax. Then one of the guys showed aces full and jumped up to reap in his pot. Ahab sprang forward with a roar and dared him to make a move for his winnings. The man did not accept the dare. He remained frozen, half-raised from his chair, arms extended toward but not touching the money.

Jerry shouted "No!" and Ahab slunk reluctantly back to the hall. When the man's color returned, Jerry assured him he'd been in no danger, but advised against sudden moves. All the men nodded with comprehension.

They began to deal the next hand. The man sitting closest to Boris was not a regular player in the game, but was a friend of a friend who'd been called at the last minute to play. He had announced when he'd arrived that he'd played in big games in Vegas. He was scared out of his wits. Every time he'd pick up his cards, Boris, who was virtually looking over the guy's shoulder, would start hooting. You could actually see the man twitch.

Boris enjoyed the whole spectacle and, unfailingly, whenever anyone would be intently studying the cards, deciding on a bet, Boo would commence his nerve-grating "hoo-hoo" kibitzing, which would usually bring Ahab to his feet with a sharp where-is-he where-is-he bark.

Needless to say, the game did not continue into the wee hours, and Jerry won a lot of money. No one ever requested to play at our apartment again.

Chapter 11

* *We Meet the Monkey People* *

Sunny days always sparked our masochism and tricked us into doing things we later regretted. On one particularly bright Sunday, we were convinced we really could handle Central Park with Boris.

As usual, the throngs descended and pelted us with questions and tried to pet or poke Boris, who as usual was not having a helluva good time. It was, after all, extremely difficult to appreciate nature when man was forever asking you to shake hands.

After about an hour and a half, as crowds came and went, we noticed that a young couple had been sitting on a bench just quietly watching us. They had spoken to each other in whispers, obviously about Boris, but neither of them had approached us. It caused us to smile at them.

When there were only two or three people standing around Boris, the couple approached shyly.

"We hate to bother you," the man said, "and we realize that you've been harassed with questions all afternoon, but could you please tell us how old your chimp is?"

"Why certainly. He's about sixteen months," I said.

"Thank you." They both cocked their heads to the side and admired Boris. "He's very beautiful."

"Thank you."

"Well, we don't want to bother you." They turned to go back to their bench.

"Oh, you're not bothering us," I said quickly, encouragingly. They were so nice and so different from the people we usually encountered that we hated to see them go. They didn't.

Their names were Carter and Kitty Howell. He was thin and blond and sold cash registers; she was ample and dark and bought wigs for Macy's. They told us that they'd always loved monkeys and apes and were seriously considering buying a chimp. We felt it our duty to dissuade them, though Boris thought otherwise. He went right to Carter and began playing with him. The man tried to conceal it, but he was visibly thrilled with Boris's attentions. From that point on, nothing we said showed any signs of affecting his ardor for an ape. We invited them back to our apartment to see

Boris on his own turf. They were so genuinely excited by our invitation that not even Jerry's unsparing description of Ahab deterred them.

Ahab lived up to his press and put on a welcoming display equal to any wild chimp's when we arrived. Kitty and Carter were oblivious to it. They had eyes only for Boris.

Boris was somehow aware of their admiration and immediately began to show off. He ran through his repertoire of chimp shticks in less than ten minutes, pulling his key from the drawer, somersaulting over the couch, banging the piano, and running through a quick bone relay with Ahab. The Howells loved it.

Then, buoyed by their enthusiasm, Boris ran amuck. He scaled our bookcase and began storming us with volumes. When we caught and scolded him, he broke free and leaped from piano bench to curtains to coffee table, leaving a wake of two spilled ashtrays, a glass of soda, and a formerly potted geranium. The Howells thought it was wonderful.

Jerry and I began to question our first impressions.

The next day, Carter phoned to invite us for dinner—with Boris!

He thought the piano was a terrific toy. Our neighbors (except for the Howells) didn't agree.

Photo: David Sagarin

I was stunned. Even our best friends and relatives had not gone so far as to include Boris in a dinner invitation. And here, virtual strangers who had seen Boris in action at his worst (at least we'd thought it was his worst) were inviting him. I asked him if he was sure he wanted us to bring Boris.

"That's why Kitty's making the dinner," he said, and I knew it was the truth.

Before leaving our apartment on the night of the dinner, I outfitted Boris in his company best, that being overalls and a matching T-shirt. He was extremely playful while I dressed him, dropping his baby oil bottle behind the couch, ruffling my hair with his feet, and I began worrying about the evening ahead, with good reason, as it turned out.

The Howells had a small but beautifully appointed apartment, a child-free, pet-free place with a glistening varnished floor and an elegant white rug. Before I sat Boris on the couch to remove his sweater, I saw at least six good reasons why our ape should not be in that apartment. One of the reasons was an extraordinary old Chinese vase on a coffee table. When I suggested that I was going to hold onto Boris, they protested.

"We want him to feel at home," Carter said. "Let him explore."

"I even bought him bananas," Kitty said. She held aloft a substantial bunch, and Boris barked his appreciation. Kitty beamed.

"Let him take one," Carter urged. I could have sworn I saw him lick his lips in anticipation.

Boris made a beeline for the bananas, and Carter and Kitty fairly drooled with pleasure as he ate one.

"Look, Carter, he loves our bananas," she said rapturously.

Love them he did. He was not only eating them, he was playing with them, squeezing the fruit out of the skin, and sucking it from his fingers.

"I think I'd better wipe his hands off," I said, "before he—" But too late. Boris wiped his own hands off, on the Howells' couch.

I was growing very uneasy. Boris was growing increasingly active. Every time Jerry or I went to stop him from touching something, the Howells would assure us that it was all right.

"Get him something to play with while we eat dinner," Carter said to Kitty.

"What does he like?" Kitty asked.

"Keys or spoons would be fine," I said.

"Oh, swell," Kitty said, and immediately whisked the dessert spoons from the table and dropped them on the rug, along with her house keys, Carter's keys, and a silver necklace.

Boris sat down to examine them while we sat down for dinner. He was playing nicely, but I was still apprehensive.

"He's having a good time," Carter assured us. "Enjoy the dinner."

I had kept a sharp eye on Boris all through the soup, but the fish required more concentration. I was just taking my second mouthful when I glanced toward Boris and almost choked.

He was standing happily bare-bottomed in the center of the room, grinding his soiled diaper into the Howells' white rug.

And the Howells didn't care! They refused to listen to an apology, and insisted we relax and let Boris enjoy himself. We protested and I cleaned the rug, but they were adamant; they wanted Boris to have and do whatever he pleased in their home. To prove it, they proceeded to shower him with amusements—watches, jewelry, canisters. When he showed an interest in the Howells' princess telephone, they put it on the floor for him. As he wrapped the cord around his neck and began to use the receiver as a mallet, the Howells gripped each other's arms excitedly.

"Look how much he likes it," they told each other.

Every time I attempted to stop Boris, Carter or Kitty stopped me.

"We want him to love it here, love us," they explained. I feared there'd be no more "here" to love if Boris kept at it. At one point, Kitty was playing with Boris on the couch and he began to pull her sweater. As she tried to ease it out of his grip, Carter shouted, "Let him rip it if it makes him happy. Don't frustrate him."

When we saw that Kitty was going along with Carter, Jerry and I stepped in. Neither of us cared about Kitty's sweater, if she didn't, but we had clothing of our own to consider. Letting Boris get away with something like that would be like urging a borderline psychotic to assert himself. It had dangerous potential.

We worried in earnest about the Howells.

Our concern was not unreasonable. They took to calling us every day to inquire about Boris. They wanted to know how he was feeling, what he was eating, what new things he did. But the phone calls were the easy part. They soon began dropping over, "just for a minute," to see Boris. When we clamped down on that, they started bearing gifts. At least three times a week, Carter and Kitty would appear at our door with "a little something" for Boris. Their little somethings ranged from toys and shoes through an identification bracelet from Tiffany's to a cashmere tennis sweater.

Jerry and I began referring to them as "the monkey people," and spent our spare time thinking up polite excuses to keep them away. We ran through a rather extensive (and, I might add, inspired) array, but still they persisted. I believe we feigned the death of my long-deceased grandfather twice.

It reached the point where politeness was a luxury beyond our mental means. We became straightforward in our no's. We were rude. But they didn't care! They appeared at our door with the regularity and staunchness of mailmen. Neither rain, snow, sleet, nor hail kept them away, and they brought relatives—Kitty's parents, Carter's cousins, their friends. Even Ahab in all his ferocious splendor did not daunt them. They were oblivious to their impositions, impervious to our rebukes. Their determination to keep in contact with Boris was awesome.

And Boo loved them, the way a cat loves a mouse, the way a hawk loves a chipmunk. They were fun prey. He could have his way with them and they never resisted. Carter encouraged Boris to jump on his chest; Kitty would simply smile a bit tighter when he yanked her hair. It was incredible, and loath as I am to admit it, we were fascinated. How many bricks could you pile atop each other before they tumbled? What, we marveled, was the Howells' breaking point?

Whatever it was, it was nowhere in sight by the end of October, because Carter and Kitty were having a gala party for Boris.

They invited at least twenty people, the cream of their crowd, and the resultant gathering was the most unhomogeneous group we'd ever encountered. No two people, barring those bound by matrimony, had anything in common, except perhaps that they were of the same species and lived within a one-hundred-fifty-mile radius of one another.

We arrived with Boris somewhere around seven (they'd called the party for six thirty, in deference to Boris's bedtime), and three-quarters of the group were already drunk.

Sensing the spirit of the occasion, Boris forgot his usual initial shyness and jumped right into things. He alternately amused and terrified the ladies with his inimitable ankle massage, and triggered locker-room memories of monkey jokes in a plumber from Brooklyn and a dentist from Queens.

Carter, deep in his cups, kept slipping Boris sips of Scotch and began talking to him like an old army buddy. Boris, like any old army buddy, accepted the camaraderie and more Scotch. When Jerry told Carter that he'd prefer him not to give Boo anything to drink, that it might upset his stomach, Carter "hoo-hoo"d.

Boris, either taking his cue from Carter or slightly inebriated himself, "hoo-hoo"d more vociferously. It was like watching the beginning of a typical bar argument that you knew would wind up in a brawl.

Kitty, a little more reasonable than Carter and considerably less drunk, attempted to disengage her husband from Boris, to whom he was clinging with intimate intensity. But while she was tugging Carter from the right and Jerry was reaching for Boris from the left, Boo, abetted by alcohol, swiftly decided that Carter was indeed his old army buddy and began huffing angrily. With bristling fur and defiantly pursed lips, he was telling everyone where to get off and being boisterously seconded by Carter.

I moved in before the situation worsened, but unfortunately after Boris had knocked Kitty's drink to the floor.

Although I knew it wasn't necessary, I apologized and offered to clean it up, but Carter had already convinced Kitty that she was to blame and she was back with a towel in seconds.

Jerry, Boris, and I attempted to work our way into a quiet corner, but met with no success. The plumber from Brooklyn had an endless stream of monkey jokes that in all earnestness he assured us we would love and, despite our protestations, he proceeded to tell them. He literally followed us around the room until an attractive woman who demonstrated Relax-a-sizers in people's homes whisked him away. She had looked at Boris earlier and told us that he was nice, but she was a cat person.

Meanwhile, Boris was growing impatient with my arms. He wriggled and dug his feet into my ribs in an attempt to free himself, but I held fast. Then he turned sulky. He no longer struggled, but he refused to look at me. Carter, seeing this, decided to play it to his advantage and coerce me into freeing Boris. He brought over two couples (one owned a bookstore on Broadway, and the other bore an esoteric preeminence for having introduced Kitty to Carter) and proceeded to enlist their aid in getting me to release Boris. After an embarrassing length of time and a reduction of their pleas to just letting Boris walk around for a minute, I agreed.

Boris, who had settled down and sort of resigned himself to my limitations, felt as if he'd gotten a reprieve. I'd no sooner loosened my hold than he leaped into the fray, snatching pretzels, nuzzling ankles, causing shrieks of delight and at least two or three more spills on the Howells' formerly snow white rug.

One young woman, who claimed to be an actress and was bedecked with more jewelry than clothes, trailed Boris around the room and attempted to imitate his knuckles-down lope. She hit her head on a table getting up and retired to the Howells' bedroom for the rest of the evening.

I began checking Jerry's watch at fifteen-minute intervals, which seemed incalculably longer, and when I saw that we'd already endured a respectable two hours, I retrieved Boris's sweater from under the sleeping actress and told Kitty and Carter we were leaving.

Carter was drunk enough to literally beg us to stay, getting down on both knees to do so. The plumber from Brooklyn thought this was hilarious and attempted to sit astride Carter's shoulders. They were in a heap on the floor when we left.

Carter and Kitty phoned the next day to tell us that the party was a smash. They asked if we were free for Halloween. When I said no, they suggested Thanksgiving. I told them we were booked up for all holidays until after New Year's. I was surprised when they didn't try to pin us down for Valentine's Day.

* *A Case of Mistaken Identity* *

Reading tales of confiscated wild pets kept us constantly aware of that terrifying possibility. It wasn't as bad as living in Nazi Germany and worrying about the Gestapo, but it wasn't as much fun as living in New York without that fear.

All of our friends, of course, knew about Boris, but we had decided it best to keep as many neighbors as possible in ignorance. Too many of them had strong negative feelings about Ahab and, frustrated because they couldn't do anything about him, would be more than pleased to nail us on Boris. So, as it happened, and, as it could only happen in New York, most of the tenants in our building were totally unaware of Boris's existence. A few months ago I ran into a former next-door neighbor who confessed she'd always thought we had a large parrot.

Among those living in unenlightenment was a rather tiny lady who occupied the apartment across the hall. We knew her only by sight, as she seemed to prefer a curious, furtive anonymity. On slow TV nights, we'd speculate on what crime she had committed, on what secrets she'd buried in her past. She rarely smiled, dressed in beatnik black, and one thought her much older than her thirty-some years. She drank Thunderbird wine and lined up the bottles in front of her door.

One evening around six, as I was preparing dinner and Jerry was out walking Ahab, I turned around to slip Boris a chunk of lettuce and was surprised to find him not there. This was unusual, for he enjoyed watching me prepare meals. It was also dangerous. With Boris, being out of sight meant getting into mischief. I called to Shep to see if Boris was in his room. When he said no, I asked him to find out what the imp was up to, and reluctantly he did.

"He's not here," Shep said, matter-of-factly. "Can I go back to my program?"

"What do you mean he's not here?"

"He's not in the kitchen, he's not in the living room, he's not in the bathroom—"

I didn't wait for him to finish. I tore down the hall, began opening all the closets, checking the windows, moving the furniture. And then I heard an eerie human mewling from the hall. I whirled and saw that the door to our apartment was ajar.

Visions of Boris climbing the elevator shaft, caught in the doors, at play in the labyrinthine depths of the basement, gripped me. (Ape mothering has a tendency to lead one's imagination to the darker sides.)

I raced to the hall and saw Boris outside the little lady's apartment across the way. The door was slightly opened and Boris was pushing her empty bottles of Thunderbird through the gap.

I have no idea of the terrors that must have haunted that poor lady, but she was moaning, "No, please, leave me alone."

Quickly I snatched Boris into my arms and, relieving him of the bottle he was clutching, placed it quietly on the doormat.

"Go home. Shave. Get out of here and take your spiders with you," she shouted.

The woman had evidently incorporated Boris's antic into a bout of delirium tremens. Hearing her shouts, Boo appeared ready to answer with a few of his own. I clamped my hand over his mouth and hurried back to the safety of our apartment, easing the door shut behind us.

Boris looked bewildered by the whole adventure. I told him to stick to Yoo-Hoo and observe the territorial imperative and he wouldn't have a thing to worry about.

Chapter 12

* *Savage Sibling* *

As Boris grew bolder, stronger, and toothier, Shep seemed to play with him less and less. Shep's earlier interest in Boo, which was primarily technical—i.e., how fast he could scale his cage, how long he could stand on his head, how far he could jump—now seemed virtually nonexistent. In fact, we noticed with concern that he was avoiding Boris.

Jerry and I discussed this one week while Shep was away, and we were determined to rectify the situation on his return. That Sunday, Shep came home in buoyant humor. He went straight to Boris's cage and was greeted with enthusiastic "hoo-hoo"s.

"Look how happy he is to see you," I pointed out.

"Yeah," Shep said, really pleased.

Boris reached out and attempted to grasp Shep's hand, but Shep was reluctant and moved back.

"He just wants to be friendly," I said, unable to conceal my annoyance. "He missed you. He wants to kiss you."

Shep looked at me. I could see I'd reached him. Then he smiled, that nice confident smile kids can give when they go along with an older and wiser parent. That melt-your-heart grin that says "You're right, mom."

With a look of new-found trust, Shep stuck out his arm and gave Boris his hand, which Boris promptly sank his teeth into.

I was shocked. Shep was screaming. Boris was doing triumphant somersaults on his shelf. I felt like the Wicked Witch of the West. I bandaged Shep's wound and assured him that he wouldn't be scarred for life, that Boris didn't hate him, but the credibility gap was palpable.

Jerry sat Shep down that night and gave him a man-to-man talk on dealing with Boris. He explained that animals respect authority and Shep had to assert himself or else Boris would think that he was the boss. Any time Boris bit, or even tried to bite, Shep was to whomp him. A coach firing up the team before a big game couldn't have been more effective. By the time Jerry had finished, Shep was convinced he could handle Boris. He was almost cocky about it.

"Why don't you let him out of the cage now?" Shep suggested.

Reasoning with our little fluff ball was not always easy.

Jerry thumped him on the back. "Sure thing."

"In fact," Shep said, "I'll let him out myself." He took the key and bent down to the padlock that secured the cage. Boris watched him with studious anticipation. "There," said Shep, and swung back the door.

Boris scampered out of the cage, lunged for Shep's leg, bit, and ran like hell. Shep's confidence was rained out in tears.

Jerry grabbed Boris and spanked him, repeating "No! No!" and holding Boo's mouth. Boris breathed his frightened "hoo-hoo"s and looked truly contrite. His beautiful brown eyes grew large and bewildered, his lips puckered in a hurtful pout, and he nuzzled and kissed Jerry with such sincere remorse that it made him wince with regret. It tugged heartstrings.

But no sooner had Jerry put Boris back down than he was back with his teeth at Shep's leg. So much for contrite chimps.

And then we really began to worry. Somehow, Boris's attack on Shep had sent our friendly, lovable, adorable, fluff-ball into a manic-destructive stage. Although Shep was standing up to him and preventing bites, Boris was completely disregarding our no's. He began ravaging our books, tearing our curtains, throwing things down from the bureau, the shelves, the cabinets, ripping his diapers, and we couldn't stop him. By the end of the week, which climaxed in the destruction of our brand-new electric coffee maker, the situation had reached crisis proportions. Jerry and I were grievously concerned.

* *Enter Saint Nick* *

Jerry called Henry Trefflich for advice. Henry gave him the phone number of Nick Carrado, the owner-trainer of Kokomo Jr., one of the world's most-talented celebrity chimps. Henry told Jerry that Nick was also a professional magician and self-defense instructor, handy sidelines when you're working with chimps.

I was impressed when Jerry phoned the office to tell me that Mr. Carrado had agreed to see him. It was like owning a fountain pen and being granted an audience with Mr. Parker.

Nick Carrado lived on West End Avenue not too far from us. When Jerry and Boris arrived, Nick greeted them at the door with a warm, fast, self-introduction. Then he took Boris in his arms, sniffed him, and handed him back to Jerry.

"He smells clean. Good. Come on in," he said. It caused Jerry to wonder whether other chimps had come to the Carrado door only to be sniffed and rejected. Nick later explained that you could tell how a chimp was being cared for by the way it smelled. He had strong feelings about people who mistreated chimps, and just meeting Nick, you knew you didn't want to be on the wrong side of his strong feelings.

Nick was one of those good-looking, powerfully rounded men whose easy smile and gentle voice belied a physical strength and emotional temper that were best left uncontested. Part of his apartment was used as a self-defense school where he taught, along with karate, jujitsu and judo, a little-known but mighty art called karrado. It incorporated the deadliest kicks, holds, throws, and blows of all the others with some Marine combat tricks thrown in. Nick had developed it himself.

The remainder of the vast apartment was decorated in total Kokomo Jr. Posters, portraits, news clippings, and assorted Kokomo Jr. publicity paraphernalia were everywhere. There were Kokomo Jr. playing cards, earrings, tie clips, and cuff links. Stationery sported Kokomo's picture, and a silk boxer's robe in the closet bore his name. It was evident that this man took his chimp seriously.

When Jerry and Boris entered the living room, Kokomo Jr. was sitting in an armchair. Boris "hoo-hoo"d cautiously from the security of Jerry's arms. Kokomo Jr. "hoo-hoo"d back without any hesitation or doubt about whose "hoo-hoo" was wearing the star.

While Boris peered around nervously, Nick jokingly explained the nature of training chimpanzees.

"It's easy," he said. "First you get a heavy two-by-four. You bring it down with all your might on the chimp's head. And then when you have his attention, you can start training him."

Nick proceeded to show Jerry and an extremely unnerved Boris what Kokomo Jr. could do. Starting with riding a bicycle and doffing his hat, Kokomo Jr. ran through paces that would put an average six-to-eight-year-old child to shame. Aside from the convoluted and hilarious stage routines that required perfect timing, Kokomo could answer yes and no with head shakes, pour his own milk and return the container to the refrigerator, eat with a fork and put his plate in the sink—and he was toilet trained!

Afterward, Nick told Kokomo to go sit in a chair and read a book, which he did, and Jerry was speechless.

"So," Nick said, settling into the couch, "you're having problems with the little guy, huh?"

Problems was hardly the word, but Jerry let it suffice. "What do you think is wrong?" Jerry asked.

"Simple," Nick said. "He's spoiled."

"Spoiled?"

"Sure. You've got to say 'Me boss, you chimp,' inside your head all the time, or else they'll run you into the ground. You don't ask a chimp to do something, you have to tell him."

"But—"

"Here. Let me show you." Nick took Boris, who immediately started screeching, and sat him down in a chair. Then he crossed Boris's arms so it looked like Boris was holding himself and said, "Fold your arms." Every time Boris tried to get away or move his arms, Nick repositioned him and told him firmly to stay. Within minutes, he had our hairy hellion sitting quietly in a chair with his arms folded. As far as Jerry was concerned, it was a miracle. According to Nick, it was only the beginning. He told Jerry he would come to our apartment that evening.

"You see," he explained later that night, before the actual workout with Boris, "it's all in the way you size up the situation and how you use your voice." He told us he had joined the Marines as a teenager and had learned there the power of an authoritative human voice. It worked with chimps, he discovered, and along with consistent discipline and lots of love, it was the way he'd turned Kokomo Jr. from a raw jungle recruit into a star.

It was time for Boris to begin basic training. Nick put Boris on the couch and folded his arms. "Now stay that way," he commanded.

Every time Boris reached out toward me and began a pleading screech, Nick would refold Boris's arms and shout, "Fold your arms!" By telling a chimp to fold his arms, he explained, you gave the animal something positive to do, as opposed to a negative *don't*. (We related this to friends who were the parents of a rather obstreperous little boy, and whenever he got out of hand after that they told him to sit on the couch and fold his arms. It worked!)

Boris howled, squealed, threw a tantrum, but continued to do whatever Nick said. And every time he did obey, Nick would rush over and kiss him, fondling him profusely. I will confess that some of Boo's more plaintive, ear-splitting shrieks did disturb us, but not as much as the Band-Aids on Shep's arms and legs. By the end of the evening, without having used anything more than a few shakes, tosses, and loud tones, "Saint" Nick Carrado had our chimp behaving like a champ.

* *Boot Camp Days* *

The real training of Boris was of course up to us. Nick had stressed the fact that the only difference between a show chimp and a pet one should be that a show chimp makes money. Giving Boris freedom to roam the house, as we had been doing, was like refusing to draw boundaries for a child. The result was a confused, rebellious animal who didn't know what was expected of him. Virtually every new thing Boris had done on his own had turned out to be a no-no, and after a while he'd just tuned us out.

Nick had also impressed us with the importance of the command "Go to your room!" (He never used the word *cage* in the presence of a chimp.)

Having read enough about the size and strength chimps attain, Jer concluded that getting Boris back into his "room" verbally was essential. He started training the following night. As with most major things—marriage, moon flights, chimp training—beginnings are the hardest.

Jer began by taking Boris's hand and bringing him to the cage door. Then, in a voice as firm as a drill sergeant's, he said, "Boris, go in your room," and he pointed.

Boris looked at Jerry's face, in the direction of Jerry's finger, instantly understood what was being asked of him—and broke for the bookcases. Jerry caught him on the second shelf and reprimanded him with a spank. Then he took Boo's hand and repeated what he'd done before. Boris attempted two more breaks, and met with the same unhappy lack of success. Half an hour later, he was going in and out of his cage happily and on command.

Saint Nick returned several nights later to see how we were coming along. Boris greeted him with a vociferous barrage of "hoo-hoo"s that suggested a certain ambivalence. When we put Boo through his paces, he didn't goof once. Nick was pleased. We were surprised, but didn't show it. Boris, it seemed, had come to terms with the lesser of two evils—us.

We asked Nick about easing the mouth-to-flesh relationship between Boris and Shep. He explained that to countenance a single bite by not punishing Boris would be the same as encouraging him. One firm but definitely unpleasant whack would save us and Boris the necessity of an endless stream of ineffectual little spanks.

As usual, Nick was right. The next time Boris tried to bite Shep, Jer slapped him across the bottom with a strap. This apparently communicated to Boris what our no-no's hadn't, that biting was out. To say that we used a strap might sound severe and had we never lived with an ape, I might think so myself. But a chimp is not a dog that you train with leash corrections nor a child with whom you can reason. A chimp is an animal that does exactly as he pleases unless checked by a superior force. Jerry's one slap with the strap saved much wear and tear on our emotions, to say nothing of Shep's flesh. Any time after that when I saw Boo approaching Shep with what I'd come to recognize as a menacing grimace, I'd say, "Be nice, Boris" in a cautionary tone, and he would. Shep, on the other hand, buoyed by the relief of not having Boris respond to him with teeth, took to working his way into Boris's heart by giving him piggyback rides. It was a bit like the school kid buying off the bully by offering him candy to preclude an argument, but it worked to their mutual delight and satisfaction. Shep would hunch down in front of Boris, who would climb aboard for a romp around the apartment, sheer rapture on his face. It would drive Ahab a little nutty, because he didn't know what was going on, but this only added to Boris's pleasure.

After several weeks of working with Boris, we were thrilled with the results and eager to show Nick what we'd accomplished. We took Boris over to Nick's apartment and he behaved perfectly. We literally beamed as Boris sat politely in a chair and neatly ate a banana.

Nick complimented us, then brought Kokomo out to show us a new routine he was working on. It was so complex, involving split-second right-and-left-hand coordination, the ability to grasp the concept of big and little, and an understanding of numbers, that it was difficult to believe I wasn't watching a midget in a fur suit—a smart midget. We returned home that evening realizing we had a long, long way to go.

* *Boris's Pajama Party* *

Nick Carrado soon became a good friend, and one weekend he even volunteered to chimp-sit Boris at his place. He had another young chimpanzee staying with Kokomo and thought it might be fun to bring them all together. We thought he was mad, and told him so, but he stuck with the invitation. His fiancée, Nancy, would be there to help pass out the bananas, and he was certain it would be no trouble at all. Nancy was an extraordinary woman who looked as if she were meant for luxury's lap but thought nothing of diapering apes and planning a home around a chimpanzee. Her courtship with Nick and Kokomo would have traumatized another female for life. Whenever I begin to think that grand passions are a thing of the past, I remember Nancy and Nick and believe again.

We packed a little bag of Pampers and clothes for Boris and dropped him off at Nick's apartment on Saturday morning. Boris looked betrayed when we deposited him in Nick's arms and waved good-bye.

"You'll have a wonderful time," I said, like any good mother leaving her child at camp. His parting glance indicated that he did not share my conviction. I pretended not to notice, and Jerry, Shep, and I took off for a carefree two days in the country.

Boris's pajama party, we learned when we returned, was a three-chimp circus. Kokomo, who looked and acted before the public as a male, was in reality a female chimpanzee whose maternal instincts blossomed with Boris. She cuddled him, carried him around, and protected him from the other little chimp, who also happened to be a female. (So much for chimp sisterhood!) Although Nick had them under control all the while, the hairy trio managed to turn his bed into a trampoline and his kitchen into a mess hall. They had grape fights, pillow fights, and relay races with Nancy's pocketbook; they flushed a variety of objects down the toilet before the bathroom was declared off-limits. Kokomo became possessive of Boris and put up a fuss about relinquishing him. Boris, wholly intimidated, deferred on all counts to his temporary foster mother. By the time we picked Boris up on Sunday night, Nancy and Nick looked a little tired. We thanked them profusely, and they told us they'd enjoyed every minute of it. Although Jerry and I didn't really doubt them, we did note that they nicely avoided the subject of future sleep-overs.

* *Down with Diapers* *

The reaction of all the apes to clothing is quite interesting.

The Ape People

Our training work with Boris was encouragingly successful; it was keeping our apartment, and Shep, whole. But it was not effective at all in dealing with a minor problem that was evolving into a major one, Boris's nudism.

Boris had developed an intense dislike for diapers. We didn't blame him; in fact, we sympathized very deeply. We would have preferred him unclothed and undiapered, but the practical reality of the situation made it impossible, which only heightened our frustration and drove us a little more whacky.

He had gone on occasional diaper strikes before, but this time he was so determined to be out of them that he tore off all his clothes and shredded them too. And when he really wanted to express his antagonism, he'd make sure he had a soiled diaper to work with before rendering it into confetti.

Sweet, calm, Rock-of-Gibraltar Thelma, who, without knowing a

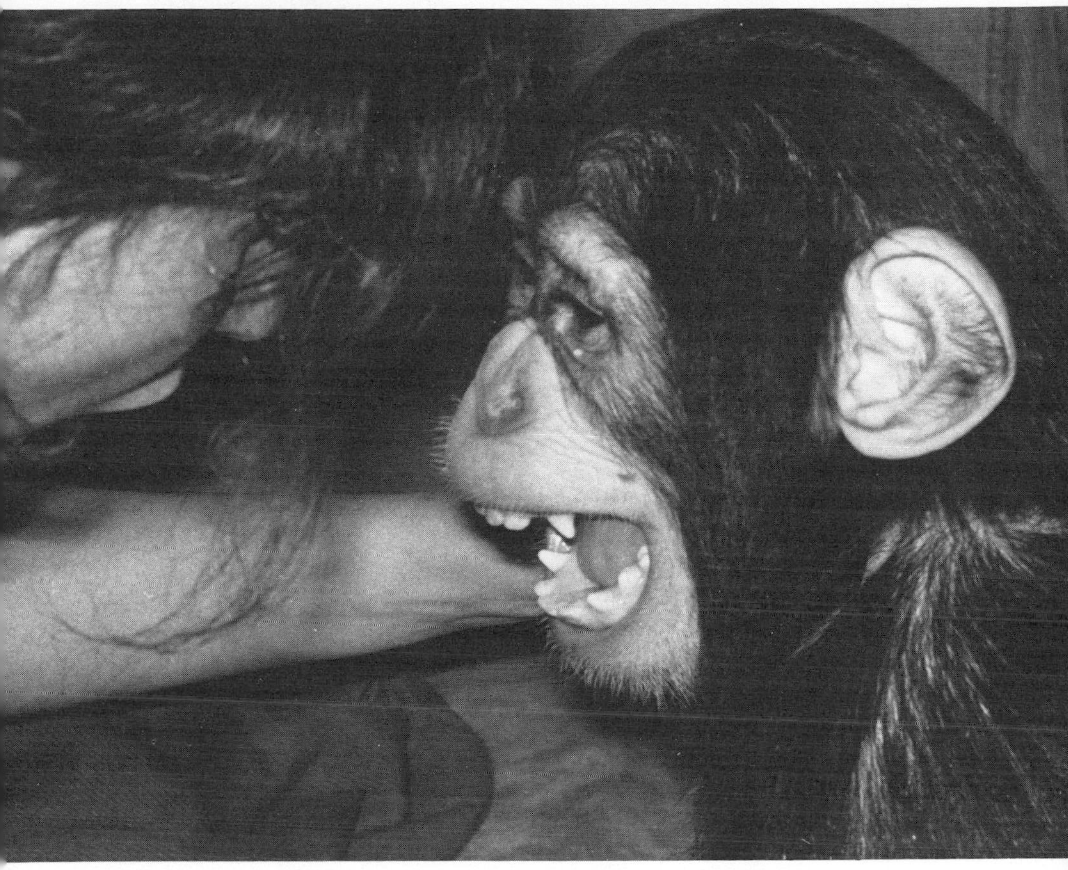

Nothing pleased Boris more than a spirited family discussion.

thing about apes or animal training had always been able to keep Boris in check, began to crumble. Every time Boo indulged in one of his hostile, antidiaper outbursts, his entire cage would have to be cleaned. When his cage began getting an hourly scrubbing, Thelma broke.

She'd never raised her voice to her little Boo-Boo before, but she raised it now, along with her hand. She spanked him soundly, gave him a tongue-lashing that would intimidate a renegade gorilla, and made him sit on a chair without moving while she cleaned up. Boris must have realized he'd gone too far, because it eliminated the problem completely for a good six months.

Chapter 13

* *Meeting the In-Laws* *

At fifteen months, Boris finally mastered drinking from a glass without spilling a drop or, as he had also been doing, dropping the glass. In fact, he became so enthusiastic over his new skill that he no longer limited his drinking to liquids. Often he would drop grapes or raisins into the glass and pour them into his mouth. He tried bread and lettuce, too, but they didn't pour as well, so he gave them up.

His mastery of the glass occurred about the same time as his mastery of the Bronx cheer. (The correlation eluded us, but the coincidence was impressive.) He'd down a glass of Yoo-Hoo and without taking a breath would wind up the feat with a triumphant razz. He liked the sound so much that he began greeting company with it instead of his usual "hoo-hoo"s. Most of our guests thought it was cute, but there were exceptions. My mother-in-law was one.

Jerry's mother had not been overly excited from the start about the addition of a jungle son to our family. She rarely asked about him in phone conversations and grew frosty whenever we bragged about his latest advancement. Although Jerry's parents lived in

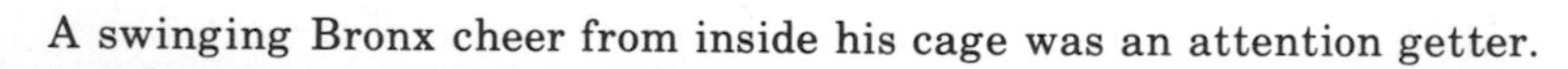
A swinging Bronx cheer from inside his cage was an attention getter.

Washington, D.C., and didn't visit often, they hadn't visited at all since we'd bought Boris. We were confident that once they got to know Boo, they would love him. Well, maybe not all that confident, but we forced ourselves to believe it in order to ease the tension that mounted as their arrival approached.

When the day came, I groomed Boris more carefully than the most indulgent ape mother. I brushed his hair until it glistened and trimmed his nails to manicured perfection, no mean task with three dextrous appendages harassing me. The end result was decidedly worth it. By the time Boris was dressed he looked terrific. How could they resist him?

My answer arrived at around five thirty. After initial hugs in the hallway and instructions on how to deal with Ahab, we ushered them into the dining room. I could see my mother-in-law withholding all judgment as she approached the cage. When she was directly in front of it, she said, "Hello, Boris."

Boris's response was his loudest, friendliest, juiciest Bronx cheer.

I shut my eyes. When I reopened them I saw my mother-in-law brushing at the front of her dress.

"He just learned to do that," I said, smiling, signaling Jerry to get the drinks. "Isn't it . . . er . . . cute?"

"I'd have preferred a drier version."

My heart sank. I led them into the living room. Jerry passed out drinks and we all drank them a little too quickly. After refills I decided it was now or never. I let Boo out of his cage.

The first thing he did was toddle over to my mother-in-law and taste her shoes. She sat rigid in her chair and said nothing until he moved on to sniff my father-in-law's. Then she said, "He's very nice." It was about as noncommittal as a grandmother could get about a grand-great ape-child, but I told myself it was a start.

Boris was on his best behavior. I don't know if he sensed our silent pleas for him to be good or if he just knew that these weren't people to monkey around with. In any case, I was grateful and lavished him with marshmallows.

As the evening wore on, Boris, that sly Lothario, began to play up to my mother-in-law, and she began to respond. She even hoisted him onto her lap several times and, to his delight and ours, petted him affectionately. I felt that my sigh of relief was audible.

When it came time to change his diaper, I decided it would be best to make it a private affair. Mom was relating warmly to Boris as our pet. I knew that for her to witness anything as prototypically human and maternal as diapering would just upset her, so I sneaked Boris off to the back bedroom. I figured I could diaper him quickly and be back before we were missed. My calculations were in error.

No sooner did we enter our bedroom than Boris became Mr. Hyde. It was as if being good for so long had just been too much for him. He raced back and forth across the bed, joyously evading my grasp. When I finally did get hold of him, he waited until I had removed his old diaper and then scampered to the bedpost and swung himself aloft. The next thing I knew, he was perched precariously on the struts of our canopied bed, determined to stay there. I climbed on the bed and managed to drag him down, but when I went to reach for a Pamper he was up again. He thought it was a game. Anything that caused me to stomp my feet and clap my hands was a game to Boris. I was beginning to look as frazzled as I felt. I climbed the bed again and got him down, this time holding him securely across the chest with my left arm while I attempted to put on his diaper with one hand. I quickly discovered that this was impossible, because I needed two hands to tear the tape with which to secure the diaper. I wasn't going to risk releasing him, so I did the first thing that came to mind, which is not to say it was either the best or most effective thing. I straddled him, pinning his arms against his sides with my knees, and proceeded to diaper him upside down.

At that moment the door to the bedroom opened. "What's taking so—" Jerry stared at me, open-mouthed. I can only imagine what I looked like by extrapolating from the expression on Jerry's face. "What the hell are you doing?" he said, and closed the door quickly behind him.

"What does it look like I'm doing?" I snapped.

"I'd hate to tell you."

"Funny. I'm only trying to change him in here so as not to upset your parents."

"A good idea if this is your new style." He held Boris while I climbed off and finished the diapering.

"The folks were getting worried," he said. "You've been in here twenty minutes. What are we going to tell them?"

I realized that the truth would only convince them that we were nuts for having Boris. I had to come up with a feasible lie. "I know," I said. "Tell them I was combing my hair, spilled a bottle of perfume, and had to clean it up."

"That sounds phony even if it were true. They'll never believe it."

"Of course they won't, but they'll suspect I'm covering up something personal, physically intimate, and won't ask again."

"Sometimes you're a genius," Jerry said.

Fortunately, it was one of those times. The remainder of the evening glided along without incident. My unexplained absence wasn't brought up again, and by the time Jerry's folks were ready to leave, Boris had them in his pink little palm. My mother-in-law even kissed him good-bye.

* *'Tis the Season to be Jolly?* *

Two weeks before Christmas, while millions across the nation happily exchanged cards with loved ones, Jerry, Shep, Thelma, and I gave each other the flu.

With Thelma unable to work and Jerry and Shep totally immobilized by soaring fever, I realized with heroic altruism that the mantle of responsibility for the entire household and Boris's health had fallen upon my own flu-aching shoulders. But did Clara Barton falter when illness struck? Of course not. So with nothing but nobility of purpose to keep me going, I survived eight horrendous days of dog walking (four times a day), food making (all times of the day and night), and Boris changing (I lost count).

The flu that year, a particularly virulent strain, had reached epidemic proportions on the East Coast. Above all else, we had to keep Boris from contracting it. The first thing I did was buy a hospital mask. I accepted the pharmacist's word that it would be effective, but secretly harbored doubts. I mean, germs were such tiny things, surely they could slip through that cheesecloth and strike, couldn't they? I tried not to think about it.

I wore the mask every time I came near Boris, which upset him wildly. He would bark at it fearfully and try to tear it from my face. Failing this, he would run and hide behind a chair and shout angry hoots at me. His distress was equaled if not surpassed by my own. I was suffocating in that damn thing. My mind staggered when I thought of doctors and nurses operating behind such masks. How on earth did they do it? After ten minutes, I'd be gasping for air. It was awful, but it did keep the germs on the human side of our family.

Just when everyone was back on their feet again, the Howells struck. They had eased off on phone calls and visits for several weeks, but we soon realized it was only because they were building up to their holiday assault. They had, they said, something special for Boris. What could we do but invite them over?

The following evening they arrived with their special present. Giggling and clutching each other excitedly, they dragged it in. *It* was enormous. It looked like a huge rubber bagel. It was, they informed us proudly, a trampoline—the perfect gift for a bouncy ape, a synthetic version of a chimpanzee's natural jungle drum. Boris took one look at it and refused to go near it; he hated it.

The Howells were visibly crushed. Jerry and I suddenly felt awful. We explained that chimpanzees were often wary of new things, that it would just take a little while for Boris to get used to the trampoline. But from the way Boris was behaving, it appeared it was going to take more than just a little while. He wouldn't even touch the thing.

We tried all sorts of inducements, such as placing marshmallows in the center, throwing his favorite ball on it, getting Shep to jump on it, but nothing worked. Boris had decided that the big rubber bagel was his enemy, and seemingly nothing we could do would change his mind.

"If it disturbs him, just throw it out," Carter said bravely, as he and Kitty put on their coats. They left looking as if they'd been betrayed.

Jerry and I worked for the remainder of the week trying to get Boris to play with the trampoline, but with no success. Carter and Kitty would phone every night and ask hopefully if Boris had jumped on it yet. When we'd tell them that he hadn't, they sounded so dejected that we took to recounting small encouragements: "He's not running away from it anymore," or "He's getting closer to touching it." But we didn't dare lie and say that the was using it, or the Howells would be over like a shot. It was one thing to depress them, but to deceive them as well would be cruel.

We left the trampoline out on our living room floor, hoping that somehow Boris would trip on it and discover what fun it was. He didn't, but Thelma did, except she didn't think it was fun at all, for she wrenched her ankle.

In desperation I tried a mean trick. Mean tricks, I'd discovered, came in handy in extreme situations. I took a bottle of Yoo-Hoo and stood in a corner with the trampoline in front of me. The only way for Boris to reach the Yoo-Hoo was to cross the trampoline.

Boris came out of his cage and ran toward me. He stopped at the side of the trampoline and reached up, beseeching me with plaintive food grunts. I held fast; it was going to be never or now. If Boo wanted a drink, he'd have to cross the rubber bagel to get it.

I waved the bottle. He "hoo-hoo"d. It was a great ape stand-off.

"Well, come and get it," I said. And without further ado, he did. He crawled up on the thing and wavered across, looking a little unsure of himself, but determined nonetheless. I picked him up and let him finish his well-earned drink. Then I put him down—right in the middle of the canvas. He looked like an actor who had accidentally stepped out in front of the footlights and decided that as long as he was there he might as well do something. He tried a little jump. Then another. In a matter of moments he was up-and-downing as efficiently as a piston. He had taken to the rubber bagel like cream cheese. I raced to the telephone to call the Howells.

They demanded to know everything: how he was jumping ("One foot or two?"), whether he was turning somersaults and bouncing off and on. I finally had to invite them over to see for themselves.

They arrived fifteen minutes later and spent the entire evening gazing rapturously at Boris enjoying their present. Whenever Boris would leave the trampoline, Carter would entice him back to it by getting down on his knees and banging the thing like a tom-tom. And every time Boris jumped on it, Kitty would give a tiny squeal, like a mouse being tickled, or scrunched. It was a Merry Christmas for the Howells after all.

* *Shooting the Works* *

Shep's ninth birthday was coming up. As far as he was concerned, it was a historic event with the import of a religious observance. At the beginning of September he'd taken to putting up daily signs in the kitchen that gave the countdown: "Only 99 more days to you-know-what!" The child was of little faith.

Once I'd committed myself to a birthday party, I realized that it was a fine opportunity for David to shoot the remaining pictures for our Boris book. David was reluctant, but he agreed. He too was of little faith, but with a lot more reason.

On the morning of Shep's party, I once again ornamented our apartment so that it looked like a wall-to-wall piñata. The decorations sparked happy memories in Boris, and he pummeled his cage exuberantly with my inflation of each new balloon. Shep was wary of Boris's enthusiasm and said so.

"If he messes things up before my friends arrive, I'll skin him."

Despite David's suggestion to let Boris get familiar with things before the shooting, I refrained from allowing him out of the cage. There was just something about the look in Shep's eyes that I didn't want to test.

David arrived at two, and Shep's friends began to drift in soon afterward. They were totally unimpressed with the fact that they were being photographed for a book, and wanted only to play with Boris. When I suggested a game of Pin-the-tail-on-the-donkey they scoffed, started to grow rowdy. One of them began to chant, "We want Boris!"

I waited five minutes, so it wouldn't look as if I was acceding to their demands, and then went to the cage. I fooled no one.

Boris bounded from his cage with the smug look of a revolutionary whose comrades had extorted his release. He knew when he had the edge. The kids screeched with excitement and the party was on. Marshmallows flew as Shep's friends vied with each other for Boris's favor. David guarded his equipment like a photographer in a war zone.

Boris darted between legs, leaped from chair to chair, and, buoyed by the cheers of his adoring audience, went bananas. He was the star and he knew it, loved it. Shep knew it too. He didn't like it a bit.

Jerry managed to calm the kids and get them to sit around the table for the cutting of the birthday cake. I got Boris to do the same. My job was easier. Boris was fascinated with the candles. He sat perfectly still as I lit them and "hoo-hoo"d softly as the flames appeared.

David's shutter clicked furiously, as if it was his last chance to get anything usable for the book. Unfortunately, it was.

The moment the kids broke into "Happy Birthday," Boris broke loose. He climbed on the table and began to huff up and down. Shep prematurely extinguished the candles, muttering something about the incendiary factors of paper tablecloths.

Boris looked as if he was working up to something I preferred not to contemplate. He seemed to be studying the kids and the goodies with equal intensity, the intensity of a kamakazi pilot. I whisked him off the table and had Jerry lock him securely in his arms while I dished out the ice cream.

David, in an uncharacteristic burst of optimism, suggested a photograph of Boris eating ice cream at the table with the children.

It was risky, but I figured Jerry or I could grab Boris before he could get into too much trouble. Again, I overestimated us and underestimated Boris, a minor miscalculation that cost us two lights on our chandelier. It happened after Boris had eaten three spoonfuls of his own ice cream, then decided that he'd prefer another bowl, one belonging to Shep's friend Michael. Michael, who'd been sitting directly across from Boris, was a nervous, excitable child, an overreactor. The moment Boris was on the table, Michael leaped from his chair, frightening Boris, who took to the high ground in true ape fashion. The high ground, in this instance, was our chandelier.

The kids booed when I said that Boris was going back in his cage, so I held him on my lap for the duration of the party. Boris didn't really object. The kids brought him grapes and marshmallows like ritual offerings. He reclined and enjoyed it. I had the feeling that somehow Boris believed he should be worshiped.

David took a few more pictures, but I could see his heart wasn't in it.

"Do you think we have a book?" I asked as he was leaving.

"Sure, but don't ask me what kind."

I didn't.

* *Notes from the Underground* *

Boris's Christmas present from us was a regulation basketball. Right from the start, he used it in ways no Harlem Globetrotter ever dreamed of. He bowled with it, sending it down the hall to crash into Ahab's pile of bones, the piano, the rubber plant. He hunted with it, bringing Shep down twice. He rode it, gnawed on it, cuddled it, and slept with it. It was more than a toy for him, it was his best friend, and he wasn't about to part with it for anything, especially not for another enforced sojourn at Trefflich's.

Because Jerry, Shep, and I were going down to Washington for a few days, Boris once again had to serve some time at Trefflich's. None of us was happy about it. Boarding Boris always filled us with guilt. I consoled myself by insisting that Boo be allowed to keep his basketball with him. Jerry was reluctant, mainly because carrying Boris and a baskeball on the subway was going to be difficult. I suggested a taxi, but the fare downtown was exorbitant and our new, tightened household budget had gone into effect the week before.

"As long as it's not rush hour, a subway will be fine," Jerry told me. "We have to start cutting down somewhere."

I feared this wasn't the place, but knew from past experience that once Jerry had committed himself to something there was no swaying him. I bundled Boris up and handed him to Jerry, along with the basketball.

"Be careful," I said.

"We're only going to the BMT, not the moon."

I would have been less concerned had it been the latter. Moon flights didn't have germs or muggers, and they were supervised, controlled. My God, anything could happen on the BMT!

Jerry managed to make it to the subway without causing much of a stir, mainly because Boris was burrowed into Jerry's coat and looked like an ordinary infant from the back. On the platform, it was a different story. The roar of the incoming train brought a shriek from Boris and his cover was blown.

Jerry tried to keep him under wraps and inconspicuous, but it wasn't possible. Boris was thrilled with the abundance of poles and was particularly delighted with the hand loops. At one point, while Jerry was holding onto a pole, Boris reached beyond and swung out on a loop. This caused a major commotion in the train, and before they even reached the next station, Jerry was bombarded with questions. He might still have made it all the way to Trefflich's if Boris had calmed down. But Boris was enjoying his subway swing. When he reached out his hand to a little old lady, she became flustered and put a quarter in it. Before Jerry could return it, Boris thanked the woman with a first-class Bronx cheer. At that point, our week-old budget went out the window and Jerry went upstairs to hail a cab.

* *New Year's Revelations* *

Since we'd made it through the year, we decided to celebrate by having a party on New Year's Eve. We were all in good spirits and good health, and it seemed just the right time to make merry.

Boris, already a veteran partygoer, got right into the preparations, particularly the edible ones. I had planned to have an enormous bowl of fruit as a centerpiece, but during the afternoon Boo whittled it down to three oranges, a Golden Delicious apple, and a couple of hard pears. I whipped up something with balloons and crepe paper at the last minute.

The guests couldn't have cared less. Boris, the wily entertainer, soon had himself eating out of the palms of their hands. They refused to allow us to put him back in his cage and insisted he join the fun. Boris was amenable. The champagne was flowing freely and Boo was working the crowd for a fair share of sips.

Shortly before midnight, I took him into the bedroom to change his diaper. As I did, I thought it might be fun to let him charge out into the living room as the infant New Year. (I'd also had a lot of champagne.) So at the stroke of midnight, when everyone was hugging and kissing and horn tooting, diaper-clad Boris tore into the party, jumped on the table, and blithely cartwheeled into the potato salad. Happy New Year!

I cleaned him off, dressed him, and told Jer to keep an eye on him while I put up some coffee. Jer, being as enthusiastic about the champagne as the rest of the guests, was not, of course, the most diligent of guardians. I didn't realize this until later, until someone told me that Boris had been stealing ice cubes from the drinks, until the first guests prepared to leave and discovered that the sleeves and pockets of their coats were soaking wet. Boris had crammed ice cubes into virtually everyone's outerwear; lots of ice cubes. It was 12 degrees outside. Our radiators could accommodate only one coat at a time. Nobody objected when I put Boo back into his cage. Nobody at all.

* *The Fan Goes Wild* *

Boris came down with pigskin fever early in the fall, and by the time of the Superbowl he was a football addict. He watched the game with breathless fascination and would hunch his shoulders and huff up and down whenever the camera trailed a player down the field. Sometimes he'd lie on his shelf popping grapes into his mouth as he watched, looking like any of a million sports fans across the nation on a Sunday afternoon. We'd gotten used to his appreciation of the game and took it for granted; if it was a tossup between the Raiders and the Vikings and an old Myrna Loy movie, Myrna Loy always lost.

On the day of the Superbowl, Jerry and Shep had gone to a friend's house to watch the game on color TV. Not being a fan myself, I turned on the TV for Boris and snuggled up on the couch with a book. All of a sudden, I heard the roar of the crowd and the announcer's excited voice describing a phenomenal run. The next thing I knew, Boris had his hands held high over his head and was clapping. I don't know whether the camera had panned to the stands and showed fans doing the same thing, or if he'd just discovered it by himself, but he was smacking his hands together with wild enthusiasm.

From that point on, clapping became part of Boris's own routine. Whenever he wanted attention, or to captivate a new visitor, or to be let out of his cage, he ran through his material like a Catskill comic. He'd open up with the Bronx cheer, follow it with folding his arms, spew out dozens of noisy kisses, and wind up with a solid round of hands-over-head clapping. It was often a continuous performance, repeated over and over until he got a food reward or freedom. It might not have been inspired, but it worked—and it impressed the hell out of our friends.

* *Believe It or Not* *

When Mr. Tourget had built Boris's cage, he told us that it would take five years for six Green Bay Packers using all their strength to weaken the structure. It took our eighteen-month-old fluff ball eleven months.

Jerry and I couldn't believe it, but it had happened. Boris's daily workouts, pulling with hands and feet against the wire mesh and tubular steel framing, had begun to separate the cage from the wall. What we'd thought for eleven months was just his way of exercising was nothing of the kind. He was deliberately trying to escape. We realized this when the frame started to give and Boris began spending his time examining the interstice between the wall and the cage, trying to thrust his hand through.

Mr. Tourget returned, mainly because he didn't believe us when we told him over the phone what had happened, and repaired the cage. "Now," he said, "it will outlast the building. But," he added, "you could put in a few extra bolts to be on the safe side."

The safe side turned out to be ten-inch carriage bolts drilled right through the limestone dining room wall into our bedroom and capped there with giant washers and nuts! It secured the cage, all right, but whenever Boris would work out against the wire, everything on our bureau would tremble ominously. It was an unnerving way to wake up.

I began to fear that the whole wall was going to come crumbling down. Jerry told me this was impossible, but Mr. Tourget had told me that what Boris had already done was impossible. I no longer believed anyone. I insisted Jerry shore up the entire cage.

He told me that I was being ridiculous. I told him to read one of my ape books. He began work on Boris's cage the next day.

I was relieved. Boris was furious. In order to reinforce the cage wall, Jerry had to remove Boris's ropes. Boo no longer cared what was added to his "room," but if something was taken away, he grew livid. He stomped and "hoo-hoo"d angrily all over the apartment as Jerry worked. When we had to put him back inside the cage before the ropes were restrung, he turned away from us and stubbornly refused to acknowledge our existence. Even after the ropes were returned, Boris sulked for days. He didn't like anyone messing with his environment, and he made sure you knew it.

Chapter 14

* *The Animal Medical Center Redux* *

Sometime around the end of January, Boris began to sneeze, a simple involuntary action that struck fear in our hearts. We dosed him with gorilla portions of Vitamin C, which he gobbled down like candy. We also set up a round-the-clock cold steam vaporizer, which kept my hair in a constant state of frizz and did something bizarre to the finish on our table, but did nothing for Boris. Despite Shep's draft detector, something that looked like an I.U.D. and buzzed when you blew on it, and a new heater, Boris came down with pneumonia again.

The doctors at the Animal Medical Center greeted him like an old friend and, even with his raging fever, Boris responded to them the same way. The trouble was, Boris was almost a year older than he'd been the last time he was in the hospital, and a year smarter. Most of the cages in the center are designed for dogs and cats; that is, they work with a simple lever latch. Boris took one look at the latch and had his cage door opened before we even said good-bye.

The doctor called two attendants, told them the problem, and they came up with some twine and a clothes hanger.

"That ought to hold him," they said.

They didn't know the hairy Houdini. Within minutes, he'd unwound the clothes hanger and was banging it against his cage. The doctor didn't wait to see what he'd do to the twine. He sent out for a padlock.

We phoned the next day to find out how Boris was doing, and the doctor was evasive.

"Well," he said, "he looks good, but I'm not really sure how he's doing."

"I don't understand," I said, growing alarmed.

"Well, ahem, his records are gone."

"Gone? How?"

"He's eaten them," he said simply.

I thought it was a pretty poor joke and was about to say so when the doctor explained that they usually keep an animal's records right on top of the cage. Boris just reached up, pulled his in, and ate them.

"Oh," was about all I could say.

The doctor told me to call back in two days. "By then," he said confidently, "we'll have something more substantial to tell you."

Two days later the doctor told me that Boris had either hidden, eaten, or tossed away the padlock key and they had to damage the cage to get him out. He added that Boris would be able to go home in three days.

"He's quite a little character," he said, with that funny, nervous laugh that usually belies a somewhat stronger sentiment.

When we came to pick Boris up, his departure was once again more of a bon voyage party than a hospital release. Boo waved, kissed, clapped his hands, and Bronx cheered all the way to the cab. If I hadn't known better, I'd have suspected him of faking pneumonia just for the attention. I wouldn't put it past him.

* Boris Gets the Munchies *

Boris's ingestion of his hospital records seemed to have opened a whole new world of edibles to him: paper. This was an especially dangerous world in the home of a writer. He didn't really want to eat the paper, but he enjoyed chewing on it, wadding it into small balls, then taking them from his mouth and playing with them. He'd pass them from hand to hand, hand to foot, and sometimes even foot to foot.

Paper towels, napkins, and toilet paper were the most readily available, and Boris indulged himself freely. He preferred *The New York Times* to *Playboy* (*The Times* wadded more easily), but nothing was sacred. He nibbled corners of envelopes, noshed on letters from friends, and totally demolished a paperback copy of *The Fountainhead.*

The situation reached our breaking point when he ate Shep's homework, a three-page report on the geography of Brazil that had included a handsome array of National Geographic pictures pasted on tasty blue construction paper. Shep had a major-felony look in his eyes when he reported the disaster to us.

"Calm down," I said. "How bad is it? Maybe we could tape it back together?"

Shep returned with something that looked as if it had been lunch for a gang of Amazon piranhas. It was beyond hope.

"It's due tomorrow," he wailed.

"Just explain to the teacher what happened," I said.

"She won't believe me. Everyone in the class gives excuses when they're late with papers. She even said she wonders what we'll come up with next."

"I'll write a note." Parental impunity.

"She won't believe you either." Filial dogmatism.

Dogmatism triumphed. Mrs. Baumgardner told Shep that the only excuses she accepted were bodily injuries that prevented students from writing, and deaths in the family. Mrs. Baumgardner was a tough cookie.

Cream cheese on a hamburger bun—again?

* *How Now, Howells?* *

Once into the renovation of Boris's cage, Jerry decided to improve it even further by adding another pole. The pole itself would be no problem, but keeping Boris away from the saw, drill, and nails would. I came up with a perfect solution: the Howells.

The Howells had often requested to be alone with Boris, and this seemed the perfect time. I phoned and asked if they'd like to watch Boris for an hour while Jerry and I fixed his cage. They were at the door in record time.

"All you have to do," I told them, "is keep Boris in our bedroom and out of trouble."

"Sure!" they said in unison. Carter fairly quivered when I plopped Boris into his arms.

I brought them to the bedroom, led them inside, and slammed the door.

For the next hour, Jer and I heard odd noises—occasional thumps and several shrieks—but ignored them all. We had a pretty good idea of what was going on, having seen some of Boris's frenetic walks on the wild side, but neither of us said anything.

When the pole was installed, Jerry and I gathered up the tools and went to the bedroom to release the Howells.

Carter was sprawled across the bed and Boris was dancing gaily upon his Adam's apple.

"Hi!" It was a very guttural greeting.

Boris "hoo-hoo"d and dashed into my arms.

Kitty was standing in the corner, smiling weakly. Her hair looked as if it had been mussed by a propeller, and she was pinning a blouse that no longer had any buttons. Carter's glasses were on the bureau, minus half the frame.

"Are you all right?" I asked.

"Sure," Carter said, but with none of his earlier enthusiasm. He looked a bit shaken as he attempted to smooth out the bed.

"What happened to your glasses?"

"Boris wanted to play with them, and he accidentally broke them."

"Why didn't you stop him?"

"We didn't want to spoil his fun," Kitty said unsteadily. She looked as if she'd just witnessed an unspeakable act and was preparing to talk about it.

"We want him to love us," Carter explained.

"But—"

Jerry pinched my arm. He was right. There was no point in trying to explain the me-boss, you-chimp cardinal rule of surviving with apes in captivity to people who refused to learn it.

But the Howells' hour with Boris must have taught them something, because the next time we saw them, Carter had contact lenses and Kitty had a headache that caused them to leave soon after arriving. The following week, she told us that a doctor had diagnosed her headaches as part of an allergic reaction to long-haired animals, and that they were buying a parakeet. Their phone calls dropped off to bimonthly chats, and their visits slowly but surely petered out. The last thing we heard was that their parakeet could peck the skin off a banana—and they'd taught it to say "Hoo-hoo-hoo"!

* *Spitting, Taxes, and Boris Strikes Again* *

When we'd first announced to our friends that we were buying a chimp, the second-most-common warning was that he'd spit. Many nights, in the months following Boris's arrival in our home, we'd reminisce about the various warnings we'd heard and laugh. The closest Boris had ever come to spitting was an inadvertently juicy Bronx cheer. Spit? Our Boo? Never! At least not until the day he ejected a grape from his mouth and became fascinated with the attendant saliva.

The next thing we knew, he was practicing producing this wondrous substance without anything in his mouth. And once he'd discovered this exciting new talent, he began to embellish it. He'd bend backward and spit toward the ceiling, lean over his shelf and spit on the floor. Soon he incorporated it into his routine: clap hands, fold arms, blow kisses, give a Bronx cheer, and spit. And as his expectorating range expanded, we learned to duck. My no's were futile, and his aim was deadly. We were forced to move the dining room table farther from the cage.

Spitting, coupled with paper chewing, soon became his number-one sport. Before long, tiny, soggy missiles were flying everywhere. I couldn't stand it. I dared not have a dinner party. I didn't even want Jerry to phone Nick Carrado, because deep inside of me I felt that somehow I'd failed.

Fortunately, Boris's spitting was a short-lived phase. After about two weeks he gave it up entirely, for no reason other than that it bored him.

Once our house was dry again, new clouds began to gather. The Internal Revenue Service had some questions to ask us about our tax return, namely about a deduction for seven hundred and fifty pounds of bananas.

Jerry had claimed Boris as a research expense, and concomitantly claimed deductions for his food, clothing, and housing. The investigator was a good sort, a gentlemen open to reason, but seven hundred and fifty pounds of bananas sounded excessive. Jerry went down to the interview with facts, figures, and photos. He gave the auditor a crash course in chimpanzees and wound up by convincing him that Boris was in no way merely a pet. His line was a master stroke of reasoning.

"I ask you," Jer said, "would any sane man keep a great ape in his dining room if it weren't for science?"

Happily, Jerry's mental state was not in question, and the auditor bought our deductions, all seven hundred and fifty pounds worth.

But with Boris, one good turn always brought forth an opposite. Just as we were feeling good about our taxes and his status as a reformed spitter, he readopted one of his old oral bad habits—biting. It happened suddenly, without warning, but not without provocation. Shep's friends were not Boris's favorite people. They would often come into the apartment and tease Boo, bang on his cage, make faces at him, do any number of the thoughtless things people do to animals, before I could stop them. This fostered no love in Boris, and his welcoming bark for children was much different from the one he gave to adults. For this reason, I limited Boris's contact with Shep's friends to just two nice kids.

But one day, Shep brought home a youngster named Peter. Peter was a frail, sallow-eyed kid who, Shep informed us, was a "semiorphan" living in a rundown Broadway hotel with an old grandfather. Peter took one look at Boris and fell in love. All he wanted to do was hold and pet him.

"Please," said Shep, "he's a semiorphan."

Peter nodded shyly and proudly. His eyes were so haunted, only a cad could refuse him.

"Well, all right. But when I let him out, don't try to grab him."

Peter gave me his word that he wouldn't. I believed him. I opened Boris's cage and the first thing he did was head straight for Peter.

Then, despite my warning, Peter lunged for him, scooped him into his arms.

Boris let out a shriek like Dracula at midnight and sank his teeth into Peter's neck.

The next shriek was Peter's, but the one following it was mine. Peter the semiorphan had just retaliated with a bite of his own!

I envisioned tabloid headlines: "CHIMP EATEN ALIVE BY EMBITTERED SEMIORPHAN" or "POOR SEMIORPHAN CHOMPED TO PIECES BY SAVAGE APE." I decided to separate them before the battle escalated. A Band-Aid fixed Peter up in minutes, but it took a week before Boris would submit to being held by a stranger again.

* *The Naked Ape* *

Those who think the first year of anything is the hardest have never lived with a chimpanzee. Although we had our problems with Boris in the first year, it wasn't until the middle of the second year that we realized when we'd been well off.

Boris had developed an independent streak, and neither punishment as we meted it out nor indulgence as we lavished it altered his course. He was determined to define himself, and clothes were not part of his image. Whether in or out of his cage, he'd strip himself naked and go on a rampage of destruction. Inside the cage, he'd merely shred his diaper and chew on bits of torn material, but outside the cage he would attack the apartment like a mad looter. When we'd attempt to stop him, he'd climb to the crawl space between the top of his cage and the ceiling and refuse to come down. From that sanctuary, he'd hoot defiantly, watch as we'd pull up a chair to climb on, and then jump to the buffet and freedom. The only time he'd obey his "Go to your cage!" command was when he was irrefutably cornered, and that wasn't often.

Our conversations with Boris regressed to one-word sentences: "No!" He wouldn't listen, wouldn't pay any attention when he was called, and would definitely not stay in clothes. He also readopted biting, and seemed amused by Shep's yelps.

Jerry was upset, I was worse, Thelma was headed for a breakdown, and Shep was going through a box of Band-Aids a week. Something had to be done.

Then one evening, Jerry said somberly, "There's only one way."

"What?" I asked apprehensively.

"A mild electric prod."

I looked at him as if he'd suggested we eat Ahab for dinner. "Never!"

"It doesn't hurt as much as it sounds like it does. It's more . . . sort of startling."

"I'll bet."

"It might bring him under control."

"So might a lobotomy."

"It was just a suggestion," Jerry snapped.

Just then, Boris dropped a bottle of twenty-five-year-old Scotch to the carpeted floor that we'd paid sixty dollars to have cleaned only three days before.

"Buy the prod," I said softly.

On Saturday Jerry went to Abercrombie & Fitch, New York's elite sportsman's paradise, and astounded the salespeople by trying out several types of electric prods, on himself! When they questioned him as to why he was attempting do-it-yourself shock treatment in their store, he explained that he would not use something on his ape that he hadn't tried out on himself first. They smiled politely and seemed happy when he left.

The prod, or magic wand, as we called it, looked a lot like a long curling iron but had two metal prongs on the end. The batteries were encased in the handle. In order for the thing to work, you had to press a button while the two metal prongs were making contact.

The next day, when Boris refused to pay attention to our no's and was all set to swing into a major household assault, Jerry used the prod.

It got Boris's attention, all right. He was in my arms like a shot.

Boris now looked at the magic wand with a certain amount of respect. Whenever he wouldn't listen, I'd simply wave the wand and without even touching him he'd be at my side. Pure magic. But that devilish imp knew that we were reluctant to use the prod, and after a few days he began turning our reluctance to his advantage. Whenever I'd brandish the prod, he'd pull the prongs right into his mouth and stare at me, actually daring me to push the button. Of course I couldn't, and right then and there he knew he had us licked.

The next thing he did was discover how to take the prod handle apart and remove the batteries. He did this many times, often hiding the batteries behind the couch. After two weeks, our magic wand was inoperable. (And I laughed when I saw *Planet of the Apes!*)

Still, Boris's behavior did improve. He no longer bit, or tried to destroy things. But nothing, absolutely nothing, could stop him from tearing off his clothes. He was determined to be a naked ape, and that was that.

I tried buying heavy denim coveralls. He went through—and I mean irreparably through—six pairs in a week. He'd either tear the straps off and then pull the garment apart, or he'd bring his feet up through the pants legs and kick the material to pieces. No matter what I bought, it would be in tatters by the end of a day, and his diaper would be shredded all over the apartment. In one week I'd spent thirty-five dollars on his clothing alone, and had nothing to show for it.

We were getting desperate. I couldn't go on purchasing five-, six-, and seven-dollar amusements daily. We called Nick Carrado, who suggested getting a dressmaker to design a heavy-duty coverall with a back zipper.

It seemed so simple a solution, we wondered why we hadn't thought of it before. But there were reasons, which we didn't learn until we tried to find someone willing and reasonable enough to design such a garment.

The first thing we learned was that there are not an awful lot of dressmakers in New York who like chimpanzees, let alone want to sew for them. Several hung up when Jerry explained our problem,

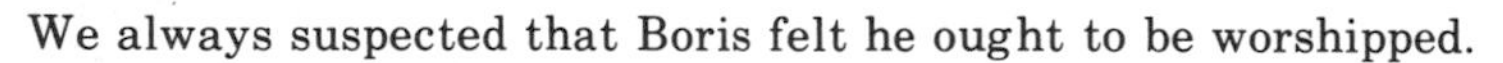

We always suspected that Boris felt he ought to be worshipped.

and most of the others were either too expensive or refused on the grounds that they didn't want to get involved. Finally, through the *Village Voice,* Jerry located a young woman who said she'd do the job if we supplied the pattern.

Neither Jerry nor I knew anything about making patterns, but necessity forced us to try. We took apart one of Boris's already torn coveralls and traced its outline on brown wrapping paper, adding a few inches here and there to provide for growth. It looked terrific to us.

Jerry phoned the young woman back and told her the pattern was ready. She gave him her address in Greenwich Village and said to stop by the following day. So, early the next morning, armed with high hopes and a definitely original design, Jerry set off downtown. He located our chimp couturiere in a run-down five-story walk-up, on the fifth floor.

The young woman, who called herself Flame, for no apparent reason, turned out to be a spaced-out hippy whose whole focus in life was paying two dollars per missive to an organization that sent astral letters to a dead swami. She nodded when Jerry showed her the pattern and promised the suit in a week.

One hundred sixty-eight hours to the second later, Jerry reappeared at her door (not that we were anxious or anything), and found, to his dismay, that Flame was freaking out. She not only didn't recall anything about Boris's suit, but she mistook Jerry for someone else and refused to let him in. After much adroit talking, Jerry managed to spark Flame's memory. She collapsed in tears into his arms, begged forgiveness for forgetting the suit, and proceeded to detail an extraordinary tale of her latest broken love affair. Jerry spent the remainder of the afternoon playing therapist and love counselor, and by the time he left, she was sitting at her sewing machine.

Three days and twenty-five dollars later, our supersuit arrived. It must have weighed five pounds! It was made of double-thick industrial denim and had a zipper on the back that looked as if it could seal a Mack truck. It was also a bit large. The extra inches here and there for growth made what were supposed to be short pants hang down below Boris's knees. When I put it on him, he nearly stooped under the weight. Yes sir, it was solid.

We put Boris back into his cage and sat down on the couch to watch what would happen. We smiled when he tried to pull off the straps, grinned when he tried to tear the material with his feet. And then suddenly the smiles and grins vanished. The seams were giving way.

I screamed, dashed for the cage, stripped Boris before he could damage the garment further, and immediately restitched every conceivable weak point. It was wasted effort. Twenty-four hours later the suit was a rag.

There was nothing left in the house for Boris to wear except some old coveralls with torn straps. Since I had no material in the house that was strong enough to serve as a strap, Jerry took one of Ahab's chain collars and fastened it on one side with several loops of thirty-pound fishing line and on the other with a tiny lock.

Boris touched the chain and stared at the lock, fingered it once or twice, then sort of shrugged. He never attempted to remove his clothes again.

The lock had done it. Somehow, the psychic effect of that little lock, a miniature version of the one that sealed his cage, convinced him he could no longer escape his garments.

Once we realized that it was the lock and chain that were keeping him in check, we sewed them on all his clothes, and normalcy returned at last.

CHAPTER 15

* *The Easter Bunny's Surprise* *

As the March winds ebbed to zephyr breezes that year, I was convinced that our stormy days with Boris were past and that we were headed for smooth sailing. Oh, I expected a squall or two now and then, but our essential Boris problems, which were food (throwing it, smearing it, and making it out of paper), clothing (destroying it), and shelter (devastating it), seemed to be gone. Boris played lovingly and toothlessly with Shep, and much more gently with Ahab. His concentration span increased and he enjoyed his own toys—basketball, hammer and pegs, trampoline—as much as he used to revel in our books. Many times he'd be content to sit in my lap and look through a magazine, getting very excited over pictures of food or television sets. In fact, things were going along so well, I couldn't understand why I felt so rotten.

Three days before Easter, a little rabbit gave me the news. I was pregnant.

Rarely in my life, except perhaps when Lassie came home in *Lassie Come Home,* had good news brought me to tears. But this was one of those times. I had read enough on the nature of maturing apes and knew enough about raising babies to realize that the two were dangerously incompatible. I had seen Boris's sibling rivalry in full flower when I'd cuddled the doll I'd brought him, and I wasn't anxious to witness it again. There was no way that Jerry and I could continue our careers and safely, fairly, raise a chimp and a baby at the same time.

Still, as impossible as we knew it was, we plunged into a frenzied search for some way to make it feasible. We searched anthropoid archives for rays of hope, consulted with veterinarians, with Nick Carrado, Henry Trefflich, and the Bronx Zoo. The answers we got were those we feared: negative. Unless we were prepared to become full-time chimp trainers, it wasn't going to work. After an agonizing month of grasping at straws, we accepted the fact that Boris would have to have a new home.

We broke the news to Shep first, and he took it badly. His pleas—"Couldn't we just keep Boris in his cage?" and "I promise I'll watch him every minute"—were heart wrenching. Thelma simply gave way to tears, daily, every time she looked at Boris. There didn't seem to be a day that went by when one of us wasn't crying. Kleenex reigned.

What was worse, Boris knew something was wrong. And what was worse than that, we knew that he knew it. Whenever he'd run through his routine that used to make us laugh, we'd cry, and he'd come over to us looking concerned and wipe the tears away with his finger, which would only bring forth new tears. My eyes began to look like taillights. After something close to forty days and forty nights, Jerry and I pulled ourselves together and began our quest for Boris's new home.

* *Placing an Ape* *

First, it was going to be nothing but the best for Boris. As long as we couldn't keep him, we were determined to ensure him the best life possible. We made up our minds immediately that we would not sell him back to Trefflich or to any other private owner, who could conceivably be forced to give him up when he reached maturity. To be torn from his mother was cruel enough. But now to be wrested from us and placed where the possibility of still another upheaval existed was out of the question. No, we were going to set Boris up for a happy, healthy, rest of his life no matter what.

Our first thought was to send him back to Sierra Leone. Apes raised in captivity are usually too humanized, too heavily imprinted by civilization, to return to the wild, but I'd hazily recalled an AP news item about a Massachusetts woman who'd taken her chimp back to Africa and spent six months on a game preserve helping it adjust to the jungle. I immediately began firing out letters to African game preserves while Jerry checked with the immigration department on what shots he and Boris would need to get there. Since I would be too pregnant to aid in a jungle orientation program, Jerry would have to do it alone. He was prepared to stay as long as necessary, and I was prepared to let him; anything for our fluff ball.

Fortunately, we were spared our marital martyrdom by a follow-up news story on hand-raised animals that had been returned to their natural environment. *Born Free* didn't always work. Many of the animals, especially simians, were unable to cope with enemies and forage on their own. The mortality rate was high. Jerry canceled his order for a pith helmet and we began to consider alternatives.

> The earliest record of a chimpanzee in a Western zoo seems to date from October 11, 1835, when one arrived at the London Zoo. It died almost at once. . . . It was nearly a century before an average chimp in an average zoo lived long enough to provide more than a nine-days', or at most a nine-months', wonder.
>
> *Zoos of the World*

Jerry and I had always loved zoos, but sending our Boris to live in one was something else. Sentencing an animal to a life behind bars was a heavy decision, and one that neither Jerry nor I was anxious to make. On the one hand, we knew that Boris would be taken care of, would have the proper food and medical treatment, would have the companionship of other chimps. On the other hand, he'd be permanently incarcerated and at the mercy of the popcorn-pelting public. Also, we had it on reputable authority that many zoos gave chimps up for research when the animals grew to obstreperous maturity, and that chilled us. We began an intensive four-month study of zoos.

What we learned was heartening. There had been extensive zoo reforms in the last twenty years. Barless caging and compatible environments had become standard in the world's mainstream zoos, and zoological research teams were continually striving to improve the captives' life-style. There were still, of course, good zoos and bad zoos. We had to separate the wheat from the chaff. We wrote letters and letters. We were not so much concerned with which zoo would want Boris, but which zoo we'd want Boris in.

We decided to start with possibilities in the United States, for selfish reasons, like being able to visit him. Although the Bronx Zoo would seem like our first choice, being geographically desirable, we didn't even consider it. Their primate lodgings were at that time emotionally and aesthetically unacceptable. And besides, their chimps looked surly.

A photo in one of our books of a thirty-acre chimpanzee enclosure being constructed at Holloman Air Force Base in New Mexico caught my eye. Thirty acres, balmy New Mexico breezes; it sounded appealing. Jerry was suspicious but phoned to find out what it was all about. The officer he spoke with was ominously uncommunicative, and was utterly silent when Jerry asked what the chimps were being brought there for. We scratched New Mexico from our list.

The chimp paddock at the Chester Zoo. *Photo: Kenneth W. Green*

The San Diego Zoological Gardens had an exemplary breeding record throughout its collection, a prime indicator of a good zoo, and the climate was ideal. Boris, we had no fear, would like it there; he had Southern California-style proclivities. But when we phoned, we learned that our enthusiasm for the zoo was not reciprocal. They informed us that their chimp population was exploding nicely and that they had no desire for any outside additions. That was a blow—Boris, blackballed at the San Diego Zoo. We resumed our quest.

The Société Royale de Zoologie d'Anvers in Antwerp interested us, and the Zoologischer Garten Basel in Switzerland, with an extraordinary gorilla-breeding record, excited us, but it was the Chester Zoo in the north of England that won our hearts.

Set upon one hundred fifty wooded, hilly acres near the west coast of England, at Upton-by-Chester, where the climate is temperate, the Chester Zoo started out as the private collection of G. S. Mottershead, a farmer with a fascination for wild animals, who kept increasing his stock until he realized that he had more of a zoo than a farm and decided to let it grow that way. The zoo opened in 1931 and was an instant success. The North of England Zoological Society was formed soon afterward, with Mr. Mottershead as secretary-director, and the farmboy's dream soon became one of the world's most beautiful and respected zoological gardens. But aside from its beauty and its reputation, and our obvious affinity for a man like Mr. Mottershead, it was the description and pictures of Chester's primate facilities that convinced us it was the perfect place for Boris. All the apes had large indoor cages and could wander out to barless, grassy, island paddocks to take the fresh air and sun at will. In *Animal Gardens,* Emily Hahn singled out the excellence of Chester's ape enclosures, saying that they lived in conditions she'd like to see copied in all zoos. And Vernon Reynolds, in *The Apes,* concurred. "The fine-looking, active, motivated gorillas, orangs, and chimpanzees at Chester make that zoo a model for the world." That cinched it. We wanted Chester. Now our only worry was, would Chester want Boris?

Jerry wrote at once, explaining our situation and our desire to find a permanent home for Boris, one in which he would not be kept singly in a small cage, in which he could socialize with others of his kind, where he would be kept forever safe from furthering the cause of medical science. We crossed our fingers and waited. Mr. Mottershead answered the following week.

We were ecstatic. Not only would Chester accept Boris, but they had at that time a group of young chimpanzees between the ages of one and three with whom they felt Boris would fit in nicely, especially since two of the group had also been hand-reared, and they were awaiting the arrival of a third from Sierra Leone! (We enjoyed thinking it would be Boris's long-lost cousin.) The letter also stated that Chester did not use chimpanzees for any stunts or training whatsoever and that Boris would live in roomy quarters with outside island accommodations and have more or less complete freedom. We could not ask for anything more.

The ensuing weeks were a blur of frenzied paperwork and intensive Boris loving. We rounded up the necessary documents for Boris's "passport"—a veterinary certificate and an export permit—made arrangements with BOAC for shipping, and cried a lot.

Boris knew something was amiss, and in the weeks and days before his departure he repeatedly tried to bring things back to normal, which scored us all the more. Whenever one of us would pet him lovingly, wistfully, in that poignantly brave but transparently sad way indicative of good-byes, he would pull away and try to strike up a game or a chase. Sometimes he'd crawl up on my lap and roll over on his back, begging to be tickled. I'd force myself to giggle and laugh with him as I had in the past, but it fooled neither of us. I found myself encouraging him and Ahab in bone relays. At least Ahab could play with him for extended periods without crying.

* *The Saddest Good-bye in the Whole World—Ever* *

We made arrangements with BOAC to fly Boris to Manchester and bought a deluxe carrying case to ship him in. Henry Trefflich had advised, and rightly so, something small and sturdy. He had explained that the less room Boris had to move around, the more unlikely his chances of being tossed about and injured, and the less he could see, the calmer he'd be. Although tranquilizers were often given to animals being shipped by air, both Henry and the Animal Medical Center warned us against them. We changed Boris's flight to one that coincided with his bedtime and confirmed with Chester that someone from the zoo would be at the airport to meet him the following morning.

The day of Boris's departure arrived, as we'd known it would, but neither Jerry, Shep, Thelma, nor I was really ready for it. I tried to be brave for Shep and Thelma's sake, and Jerry tried to be brave for mine, but none of us was doing a very good job. When Jerry told me that it was three o'clock and we had to get ready for the airport, I sincerely doubted that I'd be able to go through with it.

I dressed Boris in his red and white polo shirt. He was excited and happy, knowing that he was going out, which only tore more at my heart. The cab ride to Kennedy Airport was excruciatingly painful. Boris was loving every minute of peering out the window, and the taxi driver kept asking us why we were crying. It hurt too much to explain.

When we arrived at the BOAC cargo hangar, Boris became even more animated. As Jerry filled out the necessary forms, Boo entranced the ground crew with a nonstop round of his most engaging bits. He kissed, he shook hands, he somersaulted over and over, and made friends with everyone there. The men adored him, called him "mate" and "old chap," and seemed secretly delighted that he would soon be living on English soil. A young man from Manchester assured us that Boris would be much better off living in Chester. He told us that New York was "an 'elluva place to raise an ape."

The men offered Boris a cup of tea with milk, and when he drank it with gusto, they applauded and said he'd have no problem adjusting to British life. They had already claimed him as one of theirs.

Jerry took Boris outside for one last romp in the New York sun. It gave Boo a chance to run off some excess energy, and Jerry and I a chance to go through another package of Kleenex. And then it was time.

We went back inside the terminal and began to prepare Boris's carrying case for the flight. We took a newspaper and shredded it, just as we had when we first brought him home from Trefflich's. We put it in the case along with some lettuce, apple slices, and a piece of my sweatshirt, which he'd always cuddled at night as a sort of security blanket. Then I undressed him and Jerry put him inside.

He whimpered and thrust his arm out through one of the case windows. Jerry knelt down on the floor and held his hand. He tried to say comforting things to Boris but his voice kept cracking, so he just remained silent, holding Boris's hand. He held it for half an hour, and not once did Boris try to pull it back. He held it until they told us it was time to load Boris onto the plane. Then Jerry grabbed my hand and we kissed Boris good-bye. I could not stop the tears.

We were inconsolable, and only vaguely aware of the frightened stares people were giving us as we walked back through the main terminal. We did not watch the plane take off. It would have been more than we could endure. We needed all our strength to look at Boris's empty cage when we arrived back at our apartment.

* *A New Life* *

The next morning, we were awakened by a phone call. It was a cable from the Chester Zoo, telling us that Boris had arrived "safe, happy, and neatly packed." Jerry and I hugged each other, and for the first time since we'd made the decision to give up Boris, we were happy.

The cable was followed in a week by a long, newsy letter from G. S. Mottershead. He told us that Boris was not only doing well but was a hit with all the visitors, cutting up with antics and reveling in his increased audiences.

Photographs followed, and Jerry and I began feeling better and better. Boris's cronies looked like just the sort we'd wanted him to hang out with. The zoo informed us that the chimps in the first group he was placed with were too big and beat him up, those in the second group were too small and he beat them up. The third group, with which he was now residing, was just right.

One of the zoo's well-trained handlers eased the integration of Boris *(center)* into his peer group. *Photo: Kenneth W. Green*

It didn't take Boris *(left)* long to get into the swing of things.

Photo: Kenneth W. Green

It did take time for us to adjust to life without Boris, especially when things continued to remind us of him. Our baby, Jesse Max, wore Boo's hand-me-downs, and it was disconcerting to rediscover how easy diapering could actually be. The telephone did not ring every night with a call from the Howells. I no longer was concerned about the price of bananas. And the space in the dining room, where Boris's cage had stood, though plastered and covered with a new coat of paint, always looked as if something was missing.

But we did adjust, with no small thanks to a wonderful, salty, lion-maned gentleman named Ken Green. We struck up a pen-pal relationship with Ken, the photographer who'd taken the pictures of Boris the zoo had sent, and lucked into a wonderful friendship. Aside from being the Chester Zoo's chief photographer, Ken became Boris's personal one. The zoo allowed him out on the island to play with and photograph Boris, and his reports on Boo's accomplishments and interactions with fellow chimps eased us through the long, difficult period of adjusting to life without our fluff ball. Our gratefulness to him goes beyond words. He still visits Boris and reports to us on his progress. We can't imagine Boo's having a better friend.

* *Would We Do It All Again?* *

Often when we talk about our life with Boris, we're asked if we'd do it again. It's a curious question, and one a simple yes or no will not serve to answer. We feel privileged in having been able to know intimately and exchange love with a creature that was not born to live in civilization or with men, and selfishly we cherish that experience. But instincts inbred for thousands of years do not disappear under cute polo shirts or with a diet of cream-cheese-and-jelly sandwiches. A wild animal, be it ape or aardvark, might be able to live with man, but the cost is high. One must realize that, no matter how loving or well meaning the human, the animal is denied its natural heritage. Even allowing that the animal might receive better care than it would in the wild and might live longer, it was not the animal's choice. In order for Boris to live with us, many of his natural instincts, similar as they are to human ones, had to be forceably broken. What he would have been rewarded for in the jungle, he had to be punished for in our household. And in imposing our values, our will, our world, upon a creature that most definitely had

Chumming with Ken Green at the Chester Zoo. *Photo: Kenneth W. Green*

its own values, will, and world, there is indeed a great injustice.

Perhaps the greatest injustice happened before we even brought Boris home. Chimpanzee exploitation has gone on for many years, and the export of each baby from Africa means that two or three adult chimpanzees are wantonly sacrificed. A study done by Adriaan Kortlandt in the 1960s asserted that in Sierra Leone, Boris's birthplace, chimpanzees "are practically never heard nor seen anymore."

And though young chimps, like Boo, bear many similarities to young children, the adult chimp is an entirely different story. The mature ape develops in accordance to its genes, programed for forest survival, not apartment living.

> The juvenile ape appears adaptable and plastic and full of learning power. But it does not learn to be human or have human motives . . . while a lot of civilized customs can be trained *in*, a lot of uncivilized instincts cannot be trained *out*.
>
> *The Apes*

We were fortunate that Boris was not too humanly imprinted to readjust to life among other chimps. Had we kept him longer, the chances are that he could never have lived happily among his kind, and for that we could never have forgiven ourselves.

So would we do it all over again? No, knowing what we do now, not after having experienced the heartbreak that almost always occurs at some point in the lives of people who keep wild pets.

We will go to Chester to see Boris someday. It's unlikely that he'll remember us, but it doesn't really matter. We could never forget him.

At fifteen months, Boris adored fruit.

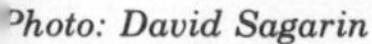

Photo: David Sagarin

At six and a half years, he still does.

Photo: Kenneth W. Green

* *Epilogue* *

Boris has been living in the Chester Zoo now for five years. During that time, he has been visited by many of our friends and relatives who've gone to England and made special pilgrimages to Chester just to see him. He's known as "the Yank chimp," and his visitors have been treated to supervised private audiences with him in the zoo's kitchen. According to the firsthand reports we've gotten, he's as engaging, outgoing, and playful as ever—though impressively larger. (Ken Green's most recent photographs put an end to all my references to "our little fluff ball.")

His keepers, two young men named Nick and Neil, adore him. They romp with him on his island, have close contact with him daily, and find him one of the friendliest chimps there. We have been told that he's not as aggressive as some of his buddies (whom I, of course, think of as bullies), but that he's more wily, managing to get whatever he wants—apple, banana, or attention—one way or another. The zoo's last communiqué informed us that he's not yet shown any indication of wanting to breed. I detected some sort of veiled implication in that but just assumed Boris hasn't met Ms. Right. Besides, I don't think I'm ready to become a grandmother.

Boris is happy at Chester. He thrives with the companionship of his kind and basks in the attentive care of his keepers. Somehow, in his own special way, he's managed to bridge the evolutionary gap that separates men from apes. He will enjoy for the rest of his days the best of all alien worlds.

Boris *(left)* hanging out with his buddies at the zoo. Photo: Kenneth W. Green